Polymer Electrolyte Membranes for Fuel Cell Devices

This volume focuses on the comprehensive discussion of the recent development toward the search for an alternative polymer electrolyte membrane (PEM), the associated challenges, and remedies to overcome the underlying practical issues. It critically assesses the merits and demerits of different PEM materials and gives a perspective on where these materials stand with respect to the standard commercial PEM. It covers PEMs, selection of membrane material, structural chemistry aspect, and recent developments of the PEM fuel cells.

Features:

- Provides a complete insight into PEMs for low-temperature fuel cells, emphasizing material selection and performance evaluation.
- Includes a comprehensive overview of the membrane types and the associated mechanism of proton conduction.
- Discusses the characterization of pertinent materials and device performance in detail.
- Reviews thermal and fluid management, transport phenomenon, degradation mechanisms, fouling and pore blocking, and tailor-made materials.
- Contains diagrammatic representation and tabular comparison in support of the discussion for a better understanding of the content.

This book is aimed at graduate students and researchers in membrane technology, fuel cells, and chemical engineering.

Polymer Electrolyte Membranes for Fuel Cell Devices

Materials, Challenges, and Future Prospects

Kamal K. Kar, Soma Banerjee, and Bablu Mordina

CRC Press
Taylor & Francis Group
Boca Raton London New York

CRC Press is an imprint of the
Taylor & Francis Group, an informa business

First edition published 2026
by CRC Press
2385 NW Executive Center Drive, Suite 320, Boca Raton FL 33431

and by CRC Press
4 Park Square, Milton Park, Abingdon, Oxon, OX14 4RN

CRC Press is an imprint of Taylor & Francis Group, LLC

ISBN: 9781032854830 (hbk)
ISBN: 9781041227359 (pbk)
ISBN: 9781003733744 (ebk)

DOI: 10.1201/9781003733744

Typeset in Times
by codeMantra

Contents

Authors' biography

Kamal K. Kar, Indian Oil Golden Jubilee Chair Professor at IIT Kanpur, is a leading academician in the area of materials/product development for supercapacitors, batteries, PEM fuel cells, water purifier, MBT-Arjuna, and artificial heart valves. He pursued higher studies at IIT Kharagpur, Iowa State University, USA, and Massachusetts Institute of Technology, USA.

Currently, he is a Professor (HAG) in the Department of Mechanical Engineering and is associated with the Interdisciplinary Programme in Materials Science and Cardiovascular and Pulmonary Diseases Research Centre, IIT Kanpur. His group made defect-engineered self-assembly to produce a three-dimensional nanohole-structured and nanoparticle-embedded porous graphene/MXene/CNT/cauliflower CNTs. He has successfully completed 50 sponsored research projects from the government and private industries. He has published more than 325 papers in international refereed journals, 160 conference papers, 11 books, 4 special issues in journals, 8 conference proceedings, 150 book chapters, and 60 granted patents, and he has more than 14000 citations. He has guided 37 doctoral students and 100 master's students so far. Currently, 8 doctoral students, 7 master's students, and a few visitors are working with him. Few of his inventions have been translated to IV, USA; ARIL, Lucknow; MEL, Chennai (MBT-Arjun Tank); SNC, Kanpur (Graphene/Nanopolymers); and IGCAR, Kalpakkam (PFBR/FBTR).

Soma Banerjee received her BSc (Chemistry Hons.), B Tech (Polymer Science and Technology), and M Tech (Polymer Science and Technology) degrees from the University of Calcutta and PhD (Materials Science) from the Indian Institute of Technology Kanpur. She has been working on the synthesis and characterization of sulfonated and phosphonated polymer composite and nanocomposite PEM for the last 15 years. Her area of research mainly focuses on the polymer electrolyte membranes for low-temperature fuel cells, polymer composites and nanocomposites, nanomaterials, bipolar plates for fuel cells, wastewater treatment, energy devices, supercapacitor, etc. She has published 18 research and review articles, 35 book chapters, presented 27 conference papers, etc., in the field of polymers and polymer nanocomposites for energy generation and storage devices. She is currently working as an Assistant Professor in the Department of Plastic Technology, School of Chemical Technology, Harcourt Butler Technical University, Kanpur.

Bablu Mordina received his BSc (Chem. Hons.) and B Tech & M Tech in Polymer Science and Technology from the University of Calcutta and PhD in Chemical Engineering from the Indian Institute of Technology Kanpur. He has 2 years of industrial experience at Kalpana Industries Ltd., Daman, on low-tension and high-tension flexible cable compounds based on polyvinyl chloride, sioplas, fire-retardant sioplas, peroxide cross-linkable polyethylene, etc. He joined the Defence Research and Development Organization (DRDO) in the year 2009 and is presently working as a Senior Scientist (Deputy Director) in the Nano Science & Coatings Division of Defence Materials and Stores Research and Development Establishment (D.M.S.R.D.E). He has 15 years of research experience in the field of polymers and polymer nanocomposites for energy devices; magnetorheological elastomers and gels for vibration damping, isolators, and actuators applications; wear-resistant polymer nanocomposites for bearing application; and nanocomposites for microwave absorption application. Apart from these, he has worked on eco-friendly coolants and high-pressure hydraulic fluids for various defence applications. He has published 20 research articles, 4 Indian patents, 5 technical reports, and 5 book chapters in reputed international journals and 5 book publishers and presented 12 invited lectures in several national and international conferences. He has been awarded DRDO Young Scientist Award in 2015, DMSRDE Technology Group Award in 2019, DMSRDE Lab Raising Day Commendation Award in 2023, National Science Day Oration Award in 2024 and International Research Awards on Civil & Environmental Engineering in 2025.

Preface

Polymer electrolyte membrane fuel cells (PEMFCs) are promising energy generation devices of the recent era. The noise-free and green technology of low-temperature fuel cells based on proton exchange membranes is grabbing the interest of researchers from interdisciplinary fields. To establish a progressive society, meeting the demand for energy is our utmost necessity. In this respect, there is a huge thrust worldwide to explore the renewable sources of energy. Hydrogen-based low-temperature fuel cells are growing over the globe, covering applications in several areas starting from portable to heavy-duty transportations. However, the performance of PEMFCs is still dependent on the costly Nafion membranes and expensive platinum catalyst. The development of alternative polymer electrolyte membranes (PEMs) that can provide proton conductivity similar to Nafion membrane at a lower cost is highly demanded for the wide acceptance of the fuel cell technology. Moreover, the commercial membrane Nafion suffers from the reduction of proton conductivity with progressive operation due to the loss of water molecules. Therefore, research should be directed to develop alternative membranes capable of withstanding long operation cycles such that the maximum benefit of it can be achieved. In this context, extensive researches are going on to develop PEMs with higher proton conductivity in order to achieve a durable PEMFC device performance.

This book provides a complete insight into the PEMs for low-temperature fuel cells with special emphasis on the material selection and performance evaluation. A comprehensive overview of the membrane types, mechanism of proton conduction, salient features, different methods of altering the conduction behavior, heat and water management inside PEM to improve the durability, mechanical properties of PEM, recent progress in this field, and future scope of the research toward the development of high-performance PEM devices have been presented. This book also discusses, in detail, the characterization of these membrane materials as well as the performance evaluation of the fuel cell devices. The potential areas of application of PEMs in different systems and subsystems are also elaborated. This book highlights the advantages and challenges and also provides effective solutions to the existing problems associated with the PEMFCs. This book will be useful to the researchers and students working in the field of materials science, energy generation, nanocomposites, and polymer science for understanding fundamentals, salient features of PEMs, and recent developments in the field of PEMFCs as well as to apply this knowledge for the development of superior PEM materials for the low-temperature fuel cells.

Chapter 1 introduces PEMs and PEMFCs. The chapter outlines different types of fuel cells with special emphasis on the component of PEMFCs, proton conduction mechanism in PEMs, fuel cell efficiency and thermodynamics of PEMs, and characteristics of PEMs. Different types of fuel cells such as proton exchange membrane fuel cells (PEMFCs), solid oxide fuel cells (SOFCs), alkaline fuel cells (AFCs), phosphoric acid fuel cells (PAFCs), molten carbonate fuel cells (MCFCs), and direct methanol fuel cells (DMFCs) are discussed in brief.

Chapter 2 discusses the different aspects of commercial Nafion membrane utilized as PEM. The proton conduction mechanism in commercial PEM is also discussed to understand the underlying phenomena. The major challenges associated with the Nafion are its high cost and loss of proton conductivity with prolonged operation due to the dehydration of membrane. This demands research for alternative PEMs. Nafion has been modified as an alternative PEM to address the issues associated with it. This chapter briefly discusses the recent developments in Nafion and modified Nafion-based composite PEMs to give a broad idea about the progress in this field.

Chapter 3 discusses the investigation of functionalized PEMs. The introduction of hydrophilic functionalized groups facilitates the water-assisted proton conduction mechanism in PEMs; however, an optimization is needed to achieve a balance between the water uptake and dimensional stability as well as mechanical integrity in the membrane. In this chapter, we have discussed sulfonated, phosphonated, fluorinated, and carboxylated PEMs along with the significance of functionalization as well as limitations of functionalization in order to achieve a balance between the hydrophilicity and mechanical strength of the membranes.

Chapter 4 emphasizes the recent advancements in anhydrous PEMs, their conduction mechanism, and recent progress in this field. The development of polybenzimidazole (PBI)-based acid-base complexes has been in the recent research trend. PBI and other polymer matrices are blended with ionic liquids to improve the proton conductivity under dehydrated conditions. Recent studies are focused on the fabrication of membrane based on polyacids, azoles, ionic liquids, metal-organic framework, and covalent organic framework anhydrous PEMs. The chapter will provide the reader with a complete overview of the different aspects of anhydrous PEMs.

Chapter 5 describes the significance of composite and nanocomposite PEMs for low-temperature fuel cell devices. The composite PEMs are fabricated by various routes such as infiltration, blending, and elctrospinning followed by activation of PEMs. Composite PEMs are fabricated with the aim to enhance the proton conductivity by the incorporation of water-retentive fillers and proton-conductive fillers. The chapter presents the recent progress in the field of composite and nanocomposite PEMs with special emphasis on the sulfonated, fluorinated, imidazole-based nanocomposite PEMs. The composite membranes at higher filler loading suffer from the problems of aggregation, leading to a reduction in PEM performance. The significance of functionalization of nanofillers to overcome the challenges of aggregation of the fillers inside the polymer matrices has also been discussed.

Chapter 6 discusses different types of non-Nafion-based PEMs, e.g., hydrocarbon-based PEM, polysulfone PEM, polyphosphazene PEM, sulfonated polyimide (SPI) PEM, polystyrene-based PEM, electrospun fiber-based PEM, crosslinked PEM, polymer blends-based PEM, natural polymer-based PEM, and microcapsules-based PEM. The swelling tendency of highly functionalized PEMs can be controlled either by crosslinking or by blending with some hydrophobic polymers. The microcapsule-based PEMs are also in recent research trend, where the proton-conducting fillers are encapsulated inside the microcapsules which are then embedded in the polymer matrix to fabricate a composite PEM with desired proton conductivity.

Chapter 7 outlines the characterization of PEMs, fabrication of membrane electrode assembly, and performance evaluation of the fuel cell devices. The membrane performances are determined by using various characterization techniques such as proton conductivity, ion exchange capacity, water and methanol uptake capacity, and thermal and mechanical properties. The in-plane and through-plane proton conductivity of the PEMs are measured in an AC impedance analyzer. The performance of the PEMFC devices is evaluated on a seven-layer membrane electrode assembly fabricated with gas diffusion layers, cathode and anode catalyst layers, current collectors, proton exchange membrane, and sealing gaskets. The device performance of the fuel cell is evaluated by analyzing the losses in the polarization curves.

Chapter 8 describes the significance of the assessment of the durability of PEMs, different modes of degradation, evaluation of membrane degradation, and thermal and water management strategies adopted for PEMFCs. The degradation assessment of PEMs is a crucial characterization for providing possible solution for long-time operation of the device without the need of frequent replacement. The evaluation is done by various test methods such as steady state test, in situ accelerated stress test, and ex situ accelerated stress test. For long-term operation without failure, a comprehensive water and thermal management is necessary. The chapter briefly explains the consequences of water flooding, fuel crossover, CO poisoning, etc., on the performance of the devices, its assessments, reasons, and possible remedies.

Finally, Chapter 9 covers the growth of PEMFCs in various sectors, e.g., portable, stationary, light-weight and heavy-weight transports. The commercial growth of this sector in the Indian and global markets is also discussed comprehensively. The future perspective of low-temperature fuel cells is also emphasized with the adaptation of advanced technologies.

The authors believe that the readers of diverse fields, e.g., materials science, polymers, nanomaterials, and energy sectors, will benefit from the comprehensive overview on PEMFCs presented in various chapters. The authors would like to extend their sincere thanks to the publishers and other authors for permitting them to reprint their images and original works. The authors hope that the book will definitely attract the attention of researchers working in the field of PEMFCs and allied sectors. In a nutshell, the authors also hope that the book will provide the students and researchers a complete understanding of the fundamentals of PEMs for low-temperature fuel cells which will definitely guide them to design and develop advanced membrane materials for the fabrication of fuel cell devices of practical importance.

Kanpur, India **Prof. Kamal K. Kar, PhD**
Kanpur, India **Dr. Soma Banerjee, PhD**
Kanpur, India **Dr. Bablu Mordina, PhD**

1 Introduction to PEM fuel cells

1.1 INTRODUCTION TO FUEL CELLS

In the present decade, a major part of the research is focused on the development of alternative energy sources [1]. In this respect, fuel cells, being the pollution-free and environmentally friendly energy-generating devices, grabbed the attention of researchers from the interdisciplinary research fields [2, 3]. Apart from polymer electrolyte membrane fuel cells (PEMFCs), researches on additional energy conversion and storage systems, such as supercapacitors [4], lithium-based batteries [5], and thermoelectric devices [6], are also capturing huge attention around the globe. Alternative energy sources are rapidly advancing to uphold environmental sustainability and economic growth, and also to fulfil geopolitical needs [7]. As global energy consumption continues to rise, dependence on non-renewable sources like coal, oil, and natural gas raises concerns due to their limited availability, rapid depletion, and harmful emissions like carbon dioxide and carbon monoxide. Fuel cells provide a promising solution to these alarming issues as they can generate energy efficiently in an environmentally friendly way. In the fuel cell devices, conversion of chemical energy occurs directly into electrical energy via an electrochemical reaction, offering a cleaner alternative to conventional combustion-based power generation systems [8]. Unlike conventional alternatives, which rely on the burning of fuels like petrol or diesel to produce energy, fuel cells produce electricity through a process that involves the reaction of fuel with oxygen, producing water and heat as byproducts in the presence of an electrolyte [9]. Hence, this process is efficient, environmentally friendly, and provides power with minimum pollutant emission. Therefore, fuel cells are key technology in the transition toward a more sustainable and low-carbon energy future. The principle behind a fuel cell is based on the movement of electrons between two electrodes, an anode and a cathode, in the presence of an electrolyte. Typically, hydrogen gas (H_2) is fed into the anode, where it is splits into protons (H^+) and electrons (e^-). The protons move through the electrolyte, while the electrons flow in the external circuit, thereby generating electricity. At the cathode, oxygen gas (O_2) is introduced, where electrons, protons, and oxygen molecules combine to form water (H_2O) as the main byproduct.

This electrochemical conversion eliminates the combustion process and drastically reduces the detrimental emissions like carbon dioxide (CO_2), carbon monoxide (CO), nitrogen oxides (NO_X), and particulate matters, making fuel cells a cleaner energy generation source. Fuel cells have the immense potential to revolutionize various sectors, comprising transportation, stationary power generation, and portable electronics. In transportation, fuel cell vehicles (FCVs) provide an efficient alternative to battery-powered electric vehicles (EVs) and diesel/petrol-powered cars. FCVs use hydrogen as fuel and release water only as the emission product, significantly reducing the carbon

DOI: 10.1201/9781003733744-1

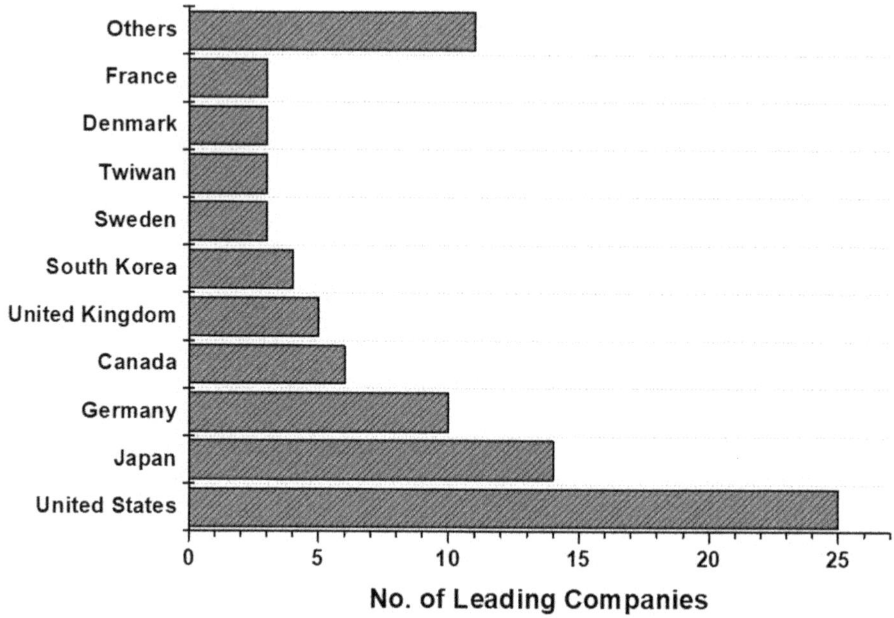

FIGURE 1.1 Major countries contributing to the development of fuel cells

footprint of transportation systems. In EVs mileage is limited by the battery's range and charging time, whereas FCVs can be refueled quickly, similar to conventional cars. This advantage makes them appealing and suitable for long-distance travel and heavy-duty uses such as buses, trucks, and trains. In the case of stationary power generation, fuel cells deliver a reliable and clean energy for both residential and industrial needs. Fuel cell systems can be utilized to generate electricity and heat for buildings or small communities, either as individual units or integrating into the existing energy grids. The ability to operate efficiently and continuously, combined with minimal environmental impact, established fuel cells as an efficient option for sustainable energy production, especially in off-grid or remote locations. Despite their many attractive advantages, the extensive acceptance of fuel cells has been hindered by several challenges. The cost of producing fuel cells, particularly the high cost of platinum used as a catalyst in many types of fuel cells, remains a significant impediment to commercialization. Additionally, the infrastructure for hydrogen manufacturing, storage, and distribution is still in its infancy, limiting the availability of hydrogen as a fuel source. However, ongoing research and continuous development in materials science, fuel cell technology, and hydrogen production methods, including green hydrogen, show great promise to decrease the cost and increase the efficiency in coming years. The major countries working on the development of fuel cells are mentioned in Figure 1.1.

1.2 TYPES OF FUEL CELLS

The fuel cells are classified on the basis of temperature, choice of electrolyte in the system, etc. The electrolyte influences the electrochemical reactions taking place at the electrode surfaces. There are numerous types of fuel cells, each of which has unique characteristics, applications, and operating conditions. PEMFCs or polymer electrolyte membrane fuel cells (also known as proton exchange membrane fuel cells), DMFCs or direct methanol fuel cells, AFCs or alkaline fuel cells, PAFCs or phosphoric acid fuel cells, MCFCs or molten carbonate fuel cells, SOFCs or solid oxide fuel cells, etc., are the typical examples of the fuel cells depending on their operating temperature, types of electrolyte, and difference in the underlying electrochemical process as depicted in Table 1.1.

a. ***Proton Exchange Membrane Fuel Cells (PEMFCs):*** Proton exchange membrane fuel cells, or polymer electrolyte membrane fuel cells, are the commonly used fuel cells, especially in transportation applications such as hydrogen-powered vehicles. They use polymer electrolyte membrane (PEM) that can conduct protons while blocking electrons and operate at low temperature (60–100°C). These cells utilize hydrogen as a fuel source and oxygen as an oxidant [10, 11]. At the anode, hydrogen gas undergoes oxidation process to obtain protons and electrons. The protons travel to the cathode via the PEM, and electrons flow via an external circuit, providing the required power to the electronics. At the cathode, protons thus generated combine with oxygen gas to produce H_2O as a byproduct. The main advantages of PEMFCs include their quick start-up, high power density, low operating temperature, high efficiency, low emissions, and suitability for mobile and remote applications. Their compact and lightweight structure make them a good choice for portable applications [11]. However, they possess certain limitations including sensitivity to impurities and high cost of production which can adversely impact their performance. Ongoing research aims to enhance the overall performance of PEMFCs by developing new and advanced materials for electrode structure, catalyst, electrolyte membrane, as well as advanced integrated system.
b. ***Solid Oxide Fuel Cells (SOFCs):*** Solid oxide fuel cells function at high temperatures, between 500°C and 1000°C. They use a non-porous solid metal oxide electrolyte (often made from Y_2O_3-stablilized ZrO_2) and can run on various types of fuels, including hydrogen, natural gas, and biogas. The anode is typically made of porous material like nickel combined with stabilized zirconia (Ni-ZrO_2), whereas cathode consists of material like lanthanum strontium manganite (Sr-$LaMnO_3$). At the anode, the fuel (H_2) undergoes oxidation, splitting into protons (H^+) and electrons (e^-). Oxygen passed from the cathode gains electrons at the electrode-electrolyte interface and is converted into oxygen ions which are then directed over the anode through diffusion into the electrolyte material. At the anode/electrolyte interface region, the oxygen ions combine with the fuel at cathode, producing water, heat, and electrons. Electrons are conducted via an external circuit, powering the electronics. The reactions for the electrochemical process are given in Eqn. (1.1–1.3) [12].

$$\text{Anodic reaction}: H_2 + O^- \rightarrow H_2O + 2e^- \tag{1.1}$$

$$\text{Cathodic reaction}: 2O_2 + 2e^- \rightarrow 4O^- \tag{1.2}$$

$$\text{Overall reaction}: H_2 + \frac{1}{2}O_2 \rightarrow H_2O \tag{1.3}$$

The SOFCs possess high efficiency and utilized for both stationary power generation and combined heat and power applications [12]. Due to their high operating temperatures, SOFCs have the ability to integrate directly with industrial processes and can produce both electricity and heat which make SOFCs highly efficient. There are other advantages, such as low emissions and options of using biomass and biogas as fuel. The main challenges include the high temperature required for operation, which results in longer start-up times and potential wear on materials, ultimately reducing their lifespan.

c. ***Alkaline Fuel Cells (AFCs):*** Alkaline fuel cells are alkaline electrolyte-based fuel cells, which typically use potassium hydroxide (KOH) as electrolyte. They work at low temperature range (between 60°C and 100°C) but are sensitive to impurities like CO_2 in the hydrogen fuel. Impurities in the fuel and supplied oxygen cause damage to the electrolyte, resulting in a decrease in the performance of AFCs. Hydrogen gas is provided to the anode side, where it dissociates into protons and electrons (Eq. 1.4). The protons pass over the electrolyte to reach the cathode. The alkaline electrolyte (KOH) allows the movement of protons from the anode to the cathode, maintaining the electrical balance. The generated electrons travel in an external circuit, creating an electrical current. At the cathode, oxygen joins with the protons and electrons to generate water (Eqn. 1.5).

$$\text{Anodic reaction}: 2H_2 + 4OH^- \rightarrow 4H_2O + 4e^- \tag{1.4}$$

$$\text{Cathodic reaction}: O_2 + 2H_2O + 4e^- \rightarrow 4OH^- \tag{1.5}$$

$$\text{Overall reaction}: 2H_2 + O_2 \rightarrow 2H_2O \tag{1.6}$$

AFCs have been used in a variety of applications, e.g., space missions by NASA and in submarines. The reason of their usage in these special strategic applications is attributed to their high efficiency and long lifespan. AFCs are less common in commercial applications due to their sensitivity to fuel impurities and the necessity for high-purity hydrogen and oxygen [13]. The AFCs offer several advantages over other fuel cell technologies, primarily due to their superior oxygen electrode kinetics and their ability to utilize a diverse range of electro-catalysts. Furthermore, AFCs can operate with high reactivity without relying on expensive noble metal catalysts for the electrodes, leading to significant cost savings. Additionally, the AFC system's radiator can be smaller as compared to the similar PEMFC system. Moreover, higher operating temperature of the alkaline fuel cell significantly enhances the performance of the system.

d. ***Phosphoric Acid Fuel Cells (PAFCs):*** PAFCs are the first fuel cells to be manufactured in large commercial quantities. They use liquid phosphoric acid as the electrolyte and operate at moderate temperatures (about 150°C to 220°C). Phosphoric acid is enclosed in a matrix made of Teflon-bonded silicon carbide. The silicon carbide matrix provides structural support and ensures that the electrolyte remains in place during operation. Teflon helps to improve the material's resistance to chemical degradation. Phosphoric acid serves as the medium for the conduction of protons (H^+). The proton conductivity of phosphoric acid is moderate, which is one of the reasons for the relatively high-operating temperatures of PAFCs. PAFCs use porous carbon as electrodes and platinum as catalyst for both the anode and cathode reactions. At the anode, hydrogen gas is fed into the cell, where it undergoes an oxidation reaction. The platinum catalyst eases the splitting of hydrogen molecules into protons (H^+) and electrons (e^-). Protons move through the phosphoric acid electrolyte to the cathode side, and electrons flow in the external circuit, producing power. At the cathode, oxygen reacts with the protons that have traveled through the electrolyte. The platinum catalyst at the cathode facilitates the reduction of oxygen to form water. The electrochemical reactions for the process are as follows (Eqn. 1.7–1.9) [14]:

$$\text{Anodic reaction}: 2H_2 \rightarrow 4H^+ + 4e^- \tag{1.7}$$

$$\text{Cathodic reaction}: O_2 + 4H^+ + 4e^- \rightarrow 2H_2O \tag{1.8}$$

$$\text{Overall reaction}: 2H_2 + O_2 \rightarrow 2H_2O \tag{1.9}$$

PAFCs have been used for stationary power generation, particularly in applications such as hospitals, commercial buildings, and backup power systems. PAFCs are less efficient than SOFCs and PEMFCs in certain applications; however, they are more durable and can tolerate impurities in the hydrogen fuel. These cells are also relatively well established in the market for stationary energy generation due to their robustness and longevity. One of the main drawbacks of PAFCs is their high cost, primarily due to the use of platinum as a catalyst, similar to PEM fuel cells. Additionally, PAFCs are often large and heavy, which limits their use to stationary power generation applications [14].

e. ***Molten Carbonate Fuel Cells (MCFCs):*** In molten carbonate fuel cells, a molten carbonate salt serves as an electrolyte combined with lithium aluminum oxide and operates at very high temperatures (600°C to 700°C) [15]. The electrochemical reactions for the MCFCs are given in the following Eqn. (1.10–1.12):

$$\text{Anodic reaction}: H_2 + CO_3^{=} \rightarrow H_2O + CO_2 + 2e^- \tag{1.10}$$

$$\text{Cathodic reaction}: 2O_2 + CO_2 + 2e^- \rightarrow CO_3^{=} \tag{1.11}$$

$$\text{Overall reaction}: H_2 + \frac{1}{2} O_2 + CO_2\,(\text{cathode}) \rightarrow 2H_2O + CO_2\,(\text{anode}) \tag{1.12}$$

These fuel cells are capable of running on a variety of fuels, including natural gas, biogas, and coal-derived fuels, making them appropriate for large-scale stationary power generation. MCFCs are of high efficiency which can reach up to 60%. When combined with heat and power systems, it may reach as high as 85%. However, the high operating temperature is essential to achieve the desirable conductivity due to the ceramic nature of the electrolyte; thus, they have a relatively slow start-up time. Additionally, similar to DMFCs, in the MCFCs external reformer is not needed. Instead, this conversion takes place within the fuel cell through a process called internal reforming, which helps lower the overall cost. The limitations of MCFCs include low durability, high operating temperature, short lifespan, and low start-up time. High operating temperatures compromise the corrosion resistance and lifespan of the cell components. Additionally, MCFCs are sensitive to sulfur which further limits their performance.

f. ***Direct Methanol Fuel Cells (DMFCs):*** Direct methanol fuel cells are similar to PEMFCs as they use proton exchange membrane as electrolyte but instead of hydrogen, methanol (CH_3OH) is supplied as the fuel. At the anode, methanol is supplied with water. Methanol undergoes oxidation, breaking down into carbon dioxide (CO_2), protons (H^+), and electrons (e^-). Electrons produced at the anode flow through an external circuit, creating an electric current. The protons (H^+) produced in the anode reaction pass over to a PEM to the cathode side. At the cathode, oxygen (O_2) from the air reacts with the protons (H^+) that have traveled through the electrolyte to produce water (H_2O) as a byproduct. The reactions involved are as follows (Eqn. 1.13–1.15) [16]:

$$\text{Anodic reaction}: CH_3OH + H_2O \rightarrow 6H^+ + CO_2 + 6e^- \tag{1.13}$$

$$\text{Cathodic reaction}: 2O_2 + 6H^+ + 6e^- \rightarrow 3H_2O \tag{1.14}$$

$$\text{Overall reaction}: CH_3OH + 2O_2 \rightarrow CO_2 + 2H_2O \tag{1.15}$$

They operate at relatively low temperature, typically between 50°C and 80°C. DMFCs are particularly promising for portable applications, such as in laptops, cell phones, and other small electronic devices. Methanol, being a liquid, can be stored and transported easily than gaseous hydrogen. Moreover, owing to the higher energy density of methanol than hydrogen, DMFCs have emerged as the convenient option for mobile power sources. The ability to extract hydrogen directly from methanol, without the need for an external reformer, helps reduce the overall cost of the fuel cell. However, DMFCs suffer from lower power density compared to hydrogen-powered fuel cells, and the methanol fuel can sometimes cause poisoning of the catalyst

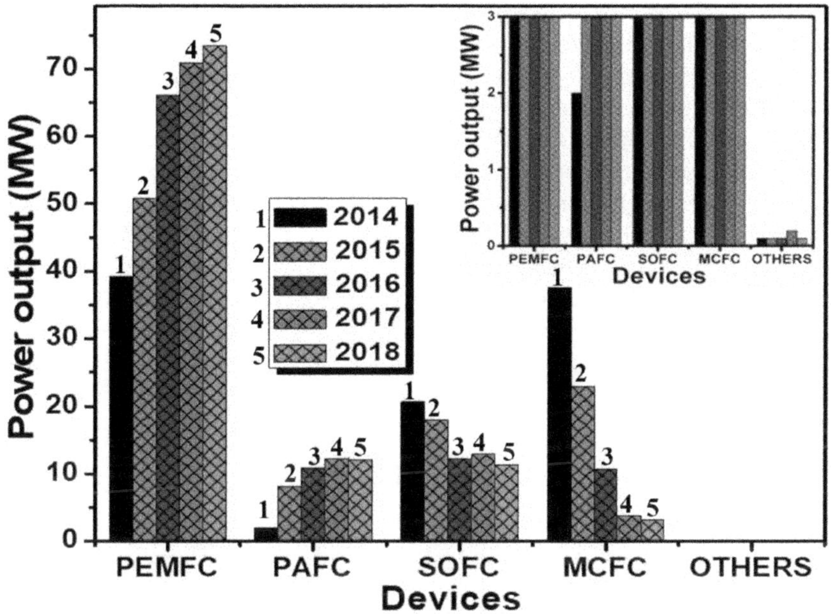

FIGURE 1.2 Output power comparisons of different devices (Data source [8])

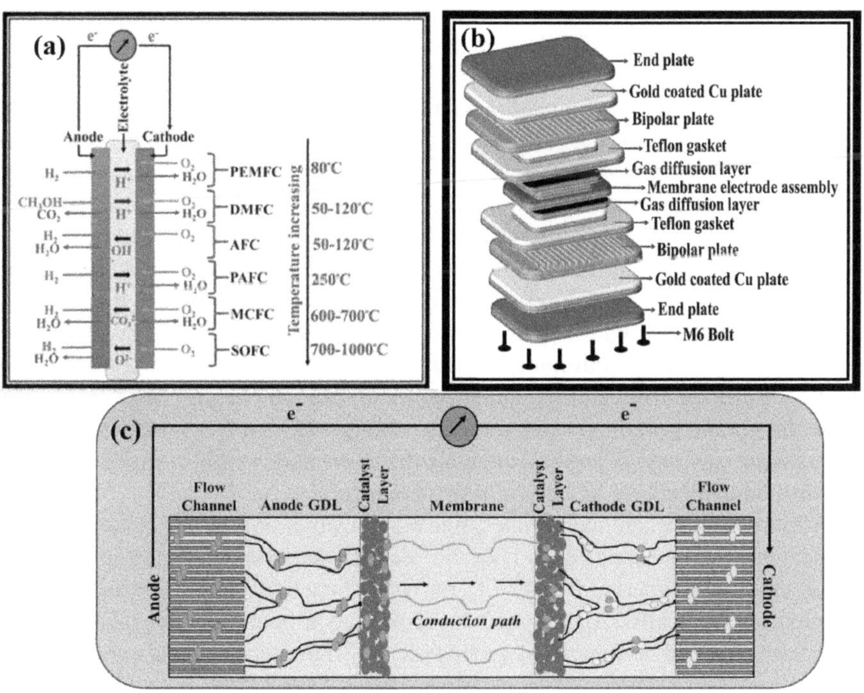

FIGURE 1.3 Schematic of (a) different types of fuel cells, (b) major components of PEM fuel cells, and (c) mechanism of operation of PEMFC

[16]. Further crossover of methanol due to the sluggish methanol oxidation reaction at the anode leads to reduced system efficiency which limits their use to low-power-consuming applications such as, camera, laptops, and other small electronic devices.

Each type of fuel cell has its own advantages and challenges, making them more suitable for specific applications. With the advancement in the technology, fuel cells are expected to serve an increasingly important role in a sustainable energy future, particularly in transportation, stationary power, and backup systems. In the subsequent sections, we will describe PEMFCs in detail. Figure 1.2 displays the power output of different types of fuel cell devices.

Different types of fuel cells have been schematically represented in Figure 1.3 with a special focus on the components and operating principle of PEMFCs. Detailed function of the components and mechanism of conduction are discussed in the succeeding sections [17].

1.3 COMPONENTS OF PEM FUEL CELLS

Proton exchange membrane fuel cells consist of several key components that work together to generate electricity from hydrogen and oxygen. These components are discussed as follows:

1.3.1 Proton exchange membrane

This is the key component of the fuel cell, made of a solid polymer electrolyte that conducts protons (positively charged hydrogen ions) while blocking electrons. It separates the anode and cathode sides of the fuel cell. In 1959, William T. Grubb introduced cation exchange membranes based on the organic network in fuel cells [18]. His efforts laid the basis of use of perfluorosulfonic acid polymer membrane as an electrolyte in now-a-days fuel cell systems. The ion exchange membrane not only provides a conductive path for protons but also separates the reactant gases. Electrically, the membrane is an insulator and thus blocks the flow of electrons from one electrode to the other through the electrolyte. In PEMFCs, the polymeric material for electrolyte is from fully fluorinated Teflon®-based family [19]. It is similar to the one used in the 1960s for space application and is manufactured by E.I. DuPont de Nemours. The most commonly used material for electrolyte membrane is commercially available from DuPont under the trade name of Nafion® No. 117. The electrolyte membrane is defined by its equivalent weight, which is inversely related to its ion exchange capacity. A typical range for the equivalent weight is given as 800 and 1100 milliequivalents for one gram of dry polymer.

Perfluorosulfonic acid membranes, like Nafion, are highly stable in harsh, adverse chemical and thermal conditions. They can resist strong bases, reducing and oxidizing acids at elevated temperature of 125°C, and offer long lifespans at selected operating conditions in fuel cells [20]. Nafion's structure has a fluoropolymer backbone with chemically attached sulfonic acid groups [21, 22]. This type of cells exhibited durability of over 50,000 hours in fuel cell tests. Dow's XUS 13204.10, a thin Nafion-like membrane with shorter side chains, displayed better interaction with water, lower electrical resistance, and exhibited good stability at high current densities; however, it has been discontinued and no longer produced [23]. Due to Nafion's high cost, alternative membranes are being developed.

1.3.2 Porous backing layer

The PEMFC uses a polymer membrane inserted between sheets of porous backing layer (also known as gas diffusion layers). This backing layer serves several functions: it helps in the diffusion of gases, provides mechanical support, creates an electrical pathway, etc. It is made from carbon-based materials, such as carbon cloth, pressed carbon fiber, or felt-like material. It helps in the evenly distribution of the gases across the catalyst surface and also allows water to pass through the membrane. The backing layer is impregnated with hydrophobic polytetrafluoroethylene (PTFE), which prevents water from pooling, thus, catalyst site can be in free contact with the gases and facilitates removal of water produced at the cathode.

An alternative design of bipolar plate was devised which incorporates porous graphite in the bipolar plates that passively manage byproduct water [24]. Liquid water from the cathode flows through the porous plate into a coolant stream, which is then directed to a reservoir. Thus, the flow of reactant stream is improved without any hurdle. Moreover, it reduces liquid water in the exhaust thereby preventing the clogging. Additionally, the flooded pores in the bipolar plate help humidify the incoming reactant gases, thus, prevents the membrane from drying at the anode. This method eliminates the need of pre-humidifying the gases and offers several advantages, such as, reduced water in the exhaust reactant streams, lower parasitic power needs, smooth operation of cell at high utilizations, smaller radiators, and a simplified control system. In dry climates, in-stack water conservation becomes crucial to maintain water balance without external water supply or refrigeration of exhaust stream.

Thermal management is also critical within the cell stack for maintaining a narrow temperature range. Proper thermal management is essential to avoid dehydration, hot spots, and dead zones within the stack by providing small temperature difference between the cell's operating temperature and ambient conditions.

1.3.3 Electrode-catalyst layer

A catalyst, usually platinum, is used on both the anode and the cathode to facilitate the chemical reactions. The catalyst helps in splitting hydrogen molecules into protons and electrons at anode, and at the cathode, it helps in the reduction of oxygen to form water. The intimate contact of catalyst particles with the electrolyte membrane facilitates the proton mobility across the membrane. The catalyst is integrated with the electrode by using a binder to form an electrode-catalyst layer. The binder not only fixes the catalyst particles at electrode but also decides the architectural structure of the electrode thereby affecting the electrochemical performance of cell.

There are two main approaches regarding the type of binder used in electrode-catalyst composition. The first approach, developed by Union Carbide and later on advanced by companies like General Electric and Dow Chemical, uses polytetrafluoroethylene (PTFE) as the binder, creating a hydrophobic, porous electrode.

The second approach focuses on the development of hydrophilic electrodes, where perfluorosulfonic acid is used as the binder. This development aims to improve membrane/catalyst contact, reducing the need for platinum [25]. In modern PEFC membrane electrode assemblies named as MEAs, catalyst is typically embedded in an electrolyte monomer solution, which helps dissolve protons and oxygen efficiently, maximizing the platinum catalyst's effectiveness. For H_2 fuel, platinum catalysts are commonly used either as pure platinum or as supported platinum catalyst on graphite or carbon at the anode. For fuels like reformate which contain CO_2, CO, N_2, etc., platinum-ruthenium alloys are preferred. The use of platinum is expensive, so efforts have been given to reduce its usage by optimizing the particle size (~3.5 nm) of platinum for optimal activity per unit mass. Additionally, alternative materials like gold or platinum alloys (Pt-transition metal) have been explored significantly, though no significant cost advantage has been achieved yet. Electrodes are typically used as thin films, either transferred to the membrane or the electrode-catalyst layer laid on the gas diffusion layer (GDL), and later on bonded to the membrane. Low platinum-catalyst-loading electrodes (total $\leq$ 1.0 mg Pt cm^{-2}) are used, achieving similar performance to higher platinum-catalyst-loading electrodes (total 2.0 to 4.0 mg Pt cm^{-2}). Recent advances in membrane electrode assembly (MEA) performance, particularly with the 7-layer MEA of 3M Company, have shown excellent durability, exceeding 10,000 hours of operation with reformate using high-speed, automated manufacturing processes [26].

1.3.4 Current Collectors

These are usually metallic components that collect and conduct the electrical current produced by the electrochemical reactions. They ensure that the current produced by the fuel cell is transferred to the external circuit. Together, these components allow PEM fuel cells to efficiently change chemical energy to electrical energy, with simultaneous generation of byproducts water and heat.

1.4 MEMBRANE—THE HEART OF PEM FUEL CELLS

The very first concept of the fuel cell was developed by Humphry Davy in 1801. Thereafter, the researches around the globe started to work in this field, and the first journal paper on the underlying reaction mechanism of fuel cells was published in 1838. PEMFC was developed by General Electric after 100 years of this event, and following this, the research on fuel cells took the momentum around the 1960s. Fuel cells are the major field of research of numerous research organizations and companies such as General Electric, NASA, Dupont, General Motor, and Toyota, for the development of fuel cell-operated stationary devices and automobile components. Figure 1.4 summarizes the major events of fuel cell research to date.

In this chapter, we will discuss the PEMFCs, also called the proton exchange membrane fuel cells, in detail. The electrolyte of these devices is a solid polymeric membrane, capable of conducting protons through it while being electrically insulating in nature. The conduction of protons through the membrane is largely dependent

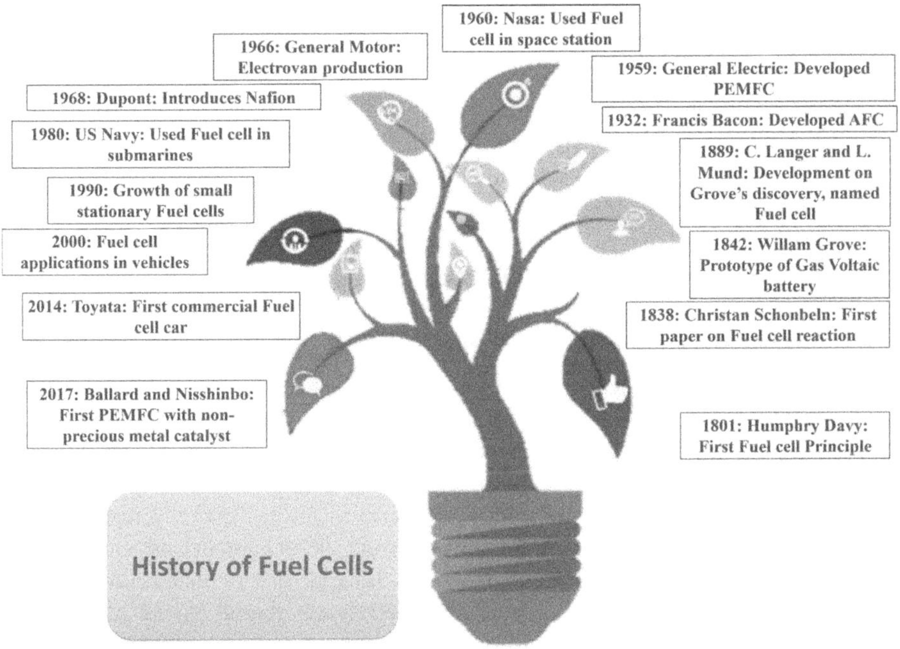

FIGURE 1.4 Growth of fuel cells (Data source from [3])

TABLE 1.1
Comparison of different fuel cells [27, 28]

Cell type	PEMFC	DMFC	AFC	PAFC	MCFC	SOFC
Electrolyte	PEM	PEM	KOH solution	Phosphoric acid	Molten K/Li carbonates	Inorganic oxides (Y_2O_3, ZrO_2)
Fuel	H_2	CH_3OH	H_2	H_2	H_2, CO, CH_4	H_2, CO, CH_4
Catalyst	Pt	Pt	Pt	Pt	Ni	Perovskites
Oxidant	O_2, air	O_2, air	O_2, air	O_2, air	CO_2, O_2, air	O_2, air
Carrier ion	H^+	H^+	OH^-	H^+	CO_3^{2-}	O^{2-}
Operating temp. (°C)	80–100	50–120	50–120	250	600–700	700–1000
Efficiency (%)	35–45	30–40	60	>40	>50–60	50–60
Capacity (kW)	<1–250	0.001–100	10–100	50–1000	<1–1000	<1–3000
Applications	Transportation, portable power	Electric utility, distributed generation	Military, space	Distributed generation	Electric utility, distributed generation	Electric utility, distributed generation

on the water content of these membranes. In addition to the PEM, the other important parts of a common PEMFC are the gas diffusion layer (GDL), catalyst layer, and bipolar plates (BPs) containing flow channels. In other words, a PEMFC can be broadly classified into two major sections: first, the membrane electrode assembly which consists of the membrane, the catalyst, and a gas diffusion layer; and second, the bipolar plates composed of flow channels, end plates, and collector plates. Figure 1.2 represents the output power comparisons of different devices with PEMFC.

1.5 PROTON CONDUCTION IN PEM

1.5.1 Water-assisted conduction mechanism in PEM

Proton transport in PEM is governed by two main mechanisms (a) Grotthuss mechanism or proton hopping, and (b) vehicle mechanism or diffusion pathways [29, 30]. In the Grotthuss mechanism, the proton mobility is assisted by the making and breaking of hydrogen bonds between the hydronium ions and water molecules. The hydronium ions themselves may exist in different forms such as $H_5O_2^+$, $H_7O_3^+$, and $H_9O_4^+$. As shown in Figure 1.5, the protons hop from one ionic site to another and conduct. The hopping from one hydrolyzing site to another is associated with local rearrangement between the proton donor and acceptor separated by a potential barrier of a few kJ/mol. This kind of mechanism is observed in a system with a strong hydrogen bond. The low potential barrier or activation energy and high proton mobility ease the conduction of protons through the membrane by the proton hopping mechanism [31].

In the vehicle mechanism, protons at first combine with the water molecules and form a complex then diffuse. The diffusion of protons is governed by a gradient in proton concentration. The diffusion process may slow down due to the bonding with other water molecules. Hence, the process is much slower as compared to the hopping mechanism. The proton conductivity of the polymer electrolyte is dependent on the state of hydration, temperature, and relative humidity (RH). At a constant

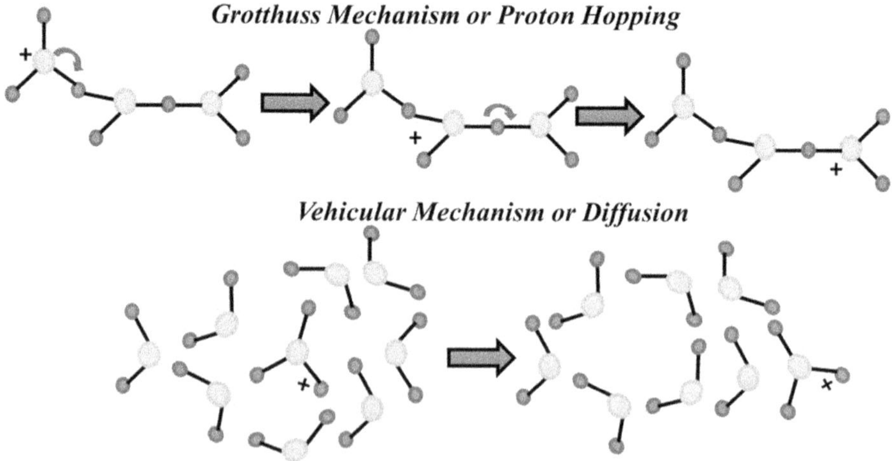

FIGURE 1.5 Proton conduction mechanisms (Concept taken from [32])

temperature, high RH, and hydrated condition, proton transport occurs preferably via the Grotthuss mechanism. Similarly, at low RH and lower proton mobility conditions, transport of the protons occurs via a diffusion pathway. Again, as the temperature is increased, hydrogen bonds begin to break, and the transport of protons takes place preferably by the vehicle mechanism.

1.5.2 Conduction at molecular level

A new gateway has been opened for understanding the proton transport phenomenon through PEMs by ab initio molecular dynamic simulations [33–35]. In hydrogen-bonded systems, fast proton exchange eliminates the existence of free protons in water. The excess protons may become a part of H_3O^+ ion, where all the three protons exist in a chemically equivalent environment. Second, the surplus proton may remain in a Zundel, $H_5O_2^+$ complex with a single proton in between two H_2O molecules, leading to the formation of a cluster. Again, the free proton in water may reside in an Eigen, $H_9O_4^+$ cluster comprising three water molecules and an H_3O^+, each strongly bound to each other. Chemically, the intermediate of all these three structures are also possible, however, these are the three clusters that appear to be dominant.

In a simplistic view, the excess protons hop from the hydronium ion to the adjacent water molecules, leaving behind a neutral water molecule. As per the Zundel and Eigen picture, H_3O^+ is a less stable molecular unit as compared to $H_5O_2^+$ complex species. The $H_5O_2^+$ acts as a basic part of the proton, and the complex with a structural defect will move together in water with its charge. A water molecule adjacent to the $H_5O_2^+$ complex fluctuates into another suitable configuration, forming a new complex with the nearest $H_5O_2^+$. Succeeding fluctuations move neighboring water molecules out of the $H_5O_2^+$ complex under consideration and move the charge by one water molecule. This kind of motion is possible with any of the remaining four protons of the $H_5O_2^+$ cluster [36].

1.6 PEM FUEL CELL EFFICIENCIES AND THERMODYNAMICS

As discussed previously, PEMFCs generate electricity by an electrochemical reaction between hydrogen and oxygen and produce water as the only byproduct. The efficiency of PEM fuel cells and the underlying thermodynamics are crucial factors in determining their performance in various applications, from portable electronics to large-scale energy systems. In this section, we will explore the efficiency and thermodynamics of PEM fuel cells in detail, including the key thermodynamic principles governing their operation, the factors influencing their efficiency, and how these cells are different from the other energy conversion technologies.

PEM fuel cells operate based on the principles of electrochemistry and thermodynamics. The primary reaction is the combination of hydrogen (H_2) and oxygen (O_2) to form water (H_2O) while releasing energy in the form of electrical power. The fundamental thermodynamic processes involved in a PEM fuel cell are governed by the Gibbs free energy, which measures the maximum work that can be done at constant temperature and pressure.

Overall reaction of PEM fuel cell can be given as follows:

The basic electrochemical reaction in a PEM fuel cell can be summarized as Eqn. (1.16):

$$2H_2 + O_2 \rightarrow 2H_2O + \text{Electrical Energy} + \text{Heat} \tag{1.16}$$

At the anode, hydrogen molecules are fragmented into protons (H^+) and electrons (e^-) as depicted in Eqn. (1.17). The cathode reactions are shown in Eqn. (1.18).

$$H_2 \rightarrow 2H^+ + 2e^- \tag{1.17}$$

$$O_2 + 4H^+ + 4e^- \rightarrow 2H_2O \tag{1.18}$$

This reaction, like any other, is governed by the laws of thermodynamics, particularly the first and second laws.

The Gibbs free energy (ΔG) for the reaction is a measure of the maximum work that can be done during the conversion of chemical energy into the electrical energy at constant temperature and pressure. It can also be calculated by using the number of electrons participated (n), Faraday's constant ($F = 96{,}487$ coulombs/mol), and potential of the cell (E).

$$W = \Delta G = -nFE \tag{1.19}$$

For the hydrogen-oxygen reaction in a PEM fuel cell, the standard Gibbs free energy change (ΔG°) at 25°C (298 K) and 1 atm pressure can be calculated by using the equations given below:

$$\Delta G = G_{f.\text{products}} - G_{f.\text{reactants}} \tag{1.20}$$

From equation (1.16),

$$\Delta G = G_{H_2O} - G_{H_2} - G_{O2} \tag{1.21}$$

$$\Delta G = -\ 237{,}180 - 0 - 0 = -\ 237{,}180 \text{ J mol}^{-1}$$

The negative value of Gibbs free energy specifies thermodynamically favorable reaction and can release energy. However, not all of the Gibbs free energy can be utilized as useful electrical work.

In ideal conditions, the maximum potential ($E \rightarrow E^\circ$) for a PEM fuel cell, without suffering any losses, may be represented as:

$$E^\circ = \frac{\Delta G}{nF} \tag{1.22}$$

$$E^\circ = \frac{237{,}180}{2 \times 96{,}485} = 1.229 \text{ V}$$

Therefore, the maximum theoretical potential of PEM fuel cells at 25°C is 1.229 V. The ideal theoretical potential at different operating temperatures may be given by the following Eqn. (1.23):

$$E_{T,P} = -\frac{\Delta G}{nF} + \frac{RT}{2F} \ln \frac{(P_{H_2})(P_{O_2})^{0.5}}{P_{H_2O}} \quad (1.23)$$

However, in real practice, this equation does not provide the value of actual working potential of fuel cell. There are various irreversible losses during the operation of the cell, such as loss of fuel, activation polarization, ohmic loss, and diffusion loss which also influence the actual working potential of the fuel cell.

The efficiency of a PEM fuel cell depends on how much of the available free energy is harnessed. The efficiency of a fuel cell is defined as the ratio of useful electrical output (work) to the total energy input (chemical energy). The maximum theoretical efficiency is determined by the reversible work, given as Gibbs free energy. However, in real systems, various losses reduce the actual efficiency.

The theoretical maximum efficiency (also known as the Carnot efficiency) for a PEM fuel cell can be derived from the relationship between Gibbs free energy (ΔG) and the enthalpy (ΔH) of the reaction. The Gibbs free energy change represents the useful electrical work, while the enthalpy change (ΔH) represents the total energy available during the reaction at conversion temperature. $T\Delta S$ is the energy unavailable due to the change in the entropy of the system.

The relationship between ΔG and ΔH is given by Eqn. 1.24:

$$\Delta G = \Delta H - T\Delta S \quad (1.24)$$

At standard conditions (25°C and 1 atm), the enthalpy change for the hydrogen-oxygen reaction is approximately 285.8 kJ/mol. Hence, the theoretical efficiency (η) of the PEM fuel cell at standard conditions is given by Eqn. 1.25.

$$\eta = \frac{\Delta G}{\Delta H} = \frac{237.18 \text{ kJ/mol}}{285.8 \text{ kJ/mol}} \approx 83\% \quad (1.25)$$

This means that under ideal conditions, a PEM fuel cell could theoretically convert up to 83% of the chemical energy of hydrogen into electricity. However, practically the efficiency is lesser since various losses are encountered.

In practice, the efficiency of PEM fuel cells is lesser due to several factors as mentioned below:

- **Activation Losses**: These losses occur at the electrodes due to the energy required to initiate the electrochemical reactions. These losses are dependent on the microstructure of the catalyst and reactant activities. They are especially significant at low current densities.

- **Ohmic Losses**: These losses arise due to the resistance of the materials (membrane, electrodes, and current collectors) to the flow of protons, electrons, and ions. Higher resistance leads to dissipation of energy as heat. The value largely depends on the geometry of cell assembly and nature of the material. The ohmic losses increase with the increase in the current density.
- **Concentration Losses**: These are caused by the depletion of reactants (hydrogen and oxygen) at the electrodes, especially at high current densities. The concentration of reactants at the electrode surface can fall below the required level, leading to the reduction in efficiency.
- **Heat Losses**: PEM fuel cells generate heat as a byproduct of the electrochemical reactions. In practice, a significant portion of the energy is lost as heat, lowering the overall efficiency.

The actual efficiency of a PEM fuel cell in practical applications typically ranges between 40% and 60% depending on the operating conditions, stack design, and stack size. During operation in combined heat and power (CHP) systems, the efficiency of PEM fuel cells can reach up to 80% or more, as the waste heat is utilized for heating purposes.

Several factors influence the efficiency of PEM fuel cells, including:

- **Operating Temperature**: The temperature at which the fuel cell operates affects the ionic conductivity of the PEM and the overall kinetics of the electrochemical reactions. Higher temperatures generally improve efficiency; however, they can also increase the wear of components.
- **Pressure**: Operating the fuel cell at higher pressures can increase the concentration of hydrogen and oxygen at the electrodes, improving the efficiency of the fuel cell. However, this also increases the complexity of the system and requires additional energy for compression.
- **Membrane Conductivity**: Efficient proton conduction through the PEM is crucial for better performance. Materials with higher proton conductivity (e.g., Nafion®) help reduce losses and improve the performance.
- **Catalyst Quality**: The efficiency of the electrochemical reactions commence at the anode and cathode depends on the quality and quantity of the catalyst material. Platinum-based catalysts are used extensively; however, high cost and limited availability are the main challenges for large-scale commercialization.

Compared to other energy conversion technologies, PEM fuel cells are highly efficient, especially in applications where clean energy and portability are key requirements. For example, internal combustion engines (ICEs) typically have thermal efficiencies in the range of 25%–30%, while PEM fuel cells can achieve an efficiency of up to 60% in electric power generation and efficiency up to 80 % utilizing the waste heat. This makes PEM fuel cells a strong candidate for applications in transportation, stationary power, and portable electronics. However, PEM fuel cells encounter challenges in terms of cost (due to the need for precious metals like platinum) and the availability of hydrogen fuel. The efficiency and practicality of PEM fuel cells will be improved as hydrogen infrastructure is developed, and alternative, less expensive catalysts replace the costly Pt catalyst.

1.7 CHARACTERISTICS OF PEM FUEL CELLS

PEM fuel cells are widely studied and used due to their efficiency, cleanliness, and versatility. Key features of the PEM fuel cells are discussed below:

1. **High Efficiency**
   PEM fuel cells are highly efficient in generating electrical energy from chemical energy. Theoretical efficiencies can reach 83%, though practical efficiencies typically range from 40% to 60%, depending on the system's design and operating conditions. Compared to conventional combustion-based systems like internal combustion engines (ICEs), PEM fuel cells generate less waste heat, making them more efficient in terms of energy conversion.
2. **Low Operating Temperature**
   PEM fuel cells function at low temperatures, between 60°C and 100°C. This is the main advantage of the PEM fuel cells in comparison to other fuel cells, like solid oxide fuel cells (SOFCs), which operate at much higher temperatures (500°C to 1,000°C). The low temperature allows for quicker startup times and eliminates the need for high-temperature insulation and complex thermal management systems.
3. **Compact and Lightweight**
   PEM fuel cells are typically compact and lightweight, making them ideal in areas where space and weight are important, such as in portable electronics, vehicles (like hydrogen-powered cars), and small-scale energy systems. They can be scaled from small, portable units to larger, stationary systems. This makes them versatile in various applications.
4. **Clean and Environmentally Friendly**
   The only byproduct of the PEM fuel cell reaction is water vapor (H_2O), which makes PEM fuel cells a clean and sustainable energy source. There are no harmful emissions, such as CO_2 or NOx, during the operation, unlike fossil fuel-based power generation. When powered by hydrogen produced from the renewable sources, PEM fuel cells can have the potential for zero greenhouse gas emissions, offering a significant advantage over traditional combustion engines.
5. **Fast Start-up Time**
   PEM fuel cells can start up quickly compared to other fuel cell types and internal combustion engines. This marks them ideal for uses that require immediate power, such as backup power systems or transportation (e.g., hydrogen-powered cars). PEM fuel cells are capable of responding rapidly to changes in load, making them suitable where the demand for power fluctuates.
6. **Low Voltage and High Current Density**
   PEM fuel cells work at low voltages (around 0.6 to 0.8 V per cell); however, they can achieve high current densities, which allows them to generate significant

amounts of power in a compact space. To generate the required voltage, individual fuel cells are stacked in series generating fuel cell stack. This makes it possible to scale up the power output as needed for different applications.

7. **Proton-Conducting Membrane**
   The membrane is the key component of PEM fuel cells, typically made from polymer electrolyte materials like Nafion®. This membrane allows protons (H^+) to pass through while blocking electrons, forcing the electrons to travel through an external circuit to generate electrical current. The performance of the PEM is highly dependent on its proton conductivity, which is influenced by temperature, humidity, and the quality of the material.
8. **Catalyst Requirements**
   The electrochemical reactions at the anode and cathode require catalysts, usually platinum or platinum-based alloys, to facilitate the splitting of hydrogen molecules and the reduction of oxygen. The high cost of platinum remains one of the impediments for widespread adoption of PEM fuel cells. Ongoing research is focused on developing cheaper and more abundant alternative catalyst material at low cost with commercial viability.
9. **Water Management**
   While water is the primary by product of the reaction, managing this water is crucial for the efficient operation of PEM fuel cells. The water produced can affect the performance of the membrane and electrodes if not properly managed. Maintaining optimal humidity levels within the fuel cell is essential for ensuring proper proton conductivity in the membrane. If the membrane dries out, its conductivity will drop which will lead to the reduction of cell performance.

1.8 CHAPTER SUMMARY

PEM fuel cells are highly efficient and promising devices for clean energy generation. The thermodynamics governing the operation, particularly the Gibbs free energy, provide a theoretical framework for understanding their efficiency. Although real-world efficiencies are lower due to losses from activation, Ohmic, concentration, and heat dissipation, PEM fuel cells still offer significant advantages over traditional energy conversion technologies, especially in terms of environmental impact and efficiency. Dedicated research and development of advanced materials, catalysts, and hydrogen infrastructure will be the key to maximize the potential of PEM fuel cells in the years to come.

REFERENCES

[1] Asghar MR, Zhang W, Su H, Zhang J, Liu H, Xing L, et al. A review of proton exchange membranes modified with inorganic nanomaterials for fuel cells. *Energy Advances* 2025;4:185–223. https://doi.org/10.1039/D4YA00446A.

[2] Kraytsberg A, Ein-Eli Y. Review of advanced materials for proton exchange membrane fuel cells. *Energy & Fuels* 2014;28:7303–30. https://doi.org/10.1021/ef501977k.

[3] Walkowiak-Kulikowska J, Wolska J, Koroniak H. Polymers application in proton exchange membranes for fuel cells (PEMFCs). *Physi Sci Rev* 2017;2:0018–51. https://doi.org/10.1515/psr-2017-0018.

[4] Kumar R, Sahoo S, Joanni E, Singh RK, Maegawa K, Tan WK, et al. Heteroatom doped graphene engineering for energy storage and conversion. *Mater Today* 2020;39:47–65. https://doi.org/10.1016/j.mattod.2020.04.010.
[5] Kumar R, Sahoo S, Joanni E, Singh RK, Tan WK, Kar KK, et al. Recent progress in the synthesis of graphene and derived materials for next generation electrodes of high performance lithium ion batteries. *Prog Energy Combust Sci* 2019;75:100786–841. https://doi.org/10.1016/j.pecs.2019.100786.
[6] Gayner C, Kar KK. Recent advances in thermoelectric materials. *Prog Mater Sci* 2016;83:330–82. https://doi.org/10.1016/j.pmatsci.2016.07.002.
[7] Manzo D, Thai R, Le HT, Venayagamoorthy GK. Fuel cell technology review: Types, economy, applications, and vehicle-to-grid scheme. *Sustainable Energy Technol. Assess* 2025;75:104229–38. https://doi.org/10.1016/j.seta.2025.104229.
[8] Banerjee S, Kar KK. Impact of degree of sulfonation on microstructure, thermal, thermomechanical and physicochemical properties of sulfonated poly ether ether ketone. *Polymer (Guildf)* 2017;109:176–86. https://doi.org/10.1016/j.polymer.2016.12.030.
[9] Patel A, Nishanth S, Patil T. Advancements in PEM fuel cell technology: Evaluating the influence of operating parameters, flow field design, bipolar plate coating, and material factors. *J Chem Technol Biotechnol* 2025;100:1159–90. https://doi.org/10.1002/jctb.7855.
[10] Homayouni H, DeVaal J, Golnaraghi F, Wang J. Voltage reduction technique for use with electrochemical impedance spectroscopy in high-voltage fuel cell and battery systems. *IEEE Trans Transp. Electrif* 2018;4:418–31. https://doi.org/10.1109/TTE.2018.2806090.
[11] Liu J, Li Q, Han Y, Zhang G, Meng X, Yu J, et al. PEMFC residual life prediction using sparse autoencoder-based deep neural network. *IEEE Trans Transp Electrif* 2019;5:1279–93. https://doi.org/10.1109/TTE.2019.2946065.
[12] Li J, Liu J. A Novel buck–boost converter with low electric stress on components. *IEEE Trans Ind Electron* 2019;66:2703–13. https://doi.org/10.1109/TIE.2018.2847646.
[13] Liu X, Li H, Wang Z. A fuel cell power conditioning system with low-frequency ripple-free input current using a control-oriented power pulsation decoupling strategy. *IEEE Trans Power Electron* 2014;29:159–69. https://doi.org/10.1109/TPEL.2013.2251004.
[14] Guo Z, Perez-Page M, Chen J, Ji Z, Holmes SM. Recent advances in phosphoric acid–based membranes for high–temperature proton exchange membrane fuel cells. *J Energy Chem* 2021;63:393–429. https://doi.org/10.1016/j.jechem.2021.06.024.
[15] Patrick JW. Handbook of fuel cells. Fundamentals technology and applications. *Fuel* 2004;83:623. https://doi.org/10.1016/j.fuel.2003.09.012.
[16] Ramasamy P, Ravindran R, Gupta N, Sardana G, Sekar I, Marthanda VA, et al. Simulation of 3D-space vector modulation for neutral point clamped inverters. *Bull Electr Eng Inf* 2023;12:719–26. https://doi.org/10.11591/eei.v12i2.4502.
[17] Sakthivel V, Kaviyarasu D, Kim AR, Kwak H Bin, Yoo DJ. Synergistic integration of sulfonated aromatic polymer blends for enhanced performance in proton exchange membrane fuel cells. *J Memb Sci* 2025;718:123641. https://doi.org/10.1016/j.memsci.2024.123641.
[18] Grubb WT, Niedrach LW. A High performance saturated hydrocarbon fuel cell. *J Electrochem Soc* 1963;110:1086–9. https://doi.org/10.1149/1.2425591.
[19] Spreafico C, Thonemann N. Prospective life cycle assessment of proton exchange membrane fuel cell. Comparing data from patents and papers. *Int J Hydrogen Energy* 2025;99:45–52. https://doi.org/10.1016/j.ijhydene.2024.12.211.
[20] Grot WGF, Munn GE, Walmsley PN. Abstract No.154. *141st National Meeting of the Electrochemical Society*, Houston, Texas: 1972.
[21] Ralph TR. Proton exchange membrane fuel cells. *Platinum Met Rev* 1997;41:102–13.
[22] Zawodzinski TA, Derouin C, Radzinski S, Sherman RJ, Smith VT, Springer TE, et al. Water uptake by and transport through nafion® 117 membranes. *J Electrochem Soc* 1993;140:1041–7. https://doi.org/10.1149/1.2056194.

[23] Prater K. The renaissance of the solid polymer fuel cell. *J Power Sources* 1990;29: 239–50. https://doi.org/10.1016/0378-7753(90)80023-7.
[24] Wheeler DJ FRYJYDPTVL. Advancement in fuel cell stack technology at International Fuel cells. 2001 Oct 1;4. *J New Mater Electrochem Syst* 2001;4:233–8.
[25] Peter MS. A preliminary investigation of radiation catalysts in fuel cells. Master of Science Thesis. Virginia Polytechnic University, 1979.
[26] All 3M Products for Fuel Cells. accessed on 29 Jun, 2025 n.d.
[27] Pachauri RK, Chauhan YK. A study, analysis and power management schemes for fuel cells. *Renewable Sustainable Energy Rev* 2015;43:1301–19. https://doi.org/10.1016/j.rser.2014.11.098.
[28] Kim DJ, Jo MJ, Nam SY. A review of polymer–nanocomposite electrolyte membranes for fuel cell application. *J Ind Eng Chem* 2015;21:36–52. https://doi.org/10.1016/j.jiec.2014.04.030.
[29] Breslau BR, Miller IF. A hydrodynamic model for electroosmosis. *Ind Eng Chem Fundam* 1971;10:554–65. https://doi.org/10.1021/i160040a003.
[30] Kreuer K-D, Paddison SJ, Spohr E, Schuster M. Transport in proton conductors for fuel-cell applications: Simulations, elementary reactions, and phenomenology. *Chem Rev* 2004;104:4637–78. https://doi.org/10.1021/cr020715f.
[31] Kreuer K-D. Proton conductivity: Materials and applications. *Chem Mater* 1996;8: 610–41. https://doi.org/10.1021/cm950192a.
[32] Gao H, Lian K. Proton-conducting polymer electrolytes and their applications in solid supercapacitors: A review. *RSC Adv* 2014;4:33091–113. https://doi.org/10.1039/C4RA05151C.
[33] Agmon N. Tetrahedral displacement: The molecular mechanism behind the debye relaxation in water. *J Phys Chem* 1996;100:1072–80. https://doi.org/10.1021/jp9516295.
[34] Schmitt UW, Voth GA. The computer simulation of proton transport in water. *J Chem Phys* 1999;111:9361–81. https://doi.org/10.1063/1.480032.
[35] Vuilleumier R, Borgis D. Transport and spectroscopy of the hydrated proton: A molecular dynamics study. *J Chem Phys* 1999;111:4251–66. https://doi.org/10.1063/1.479723.
[36] Eikerling M, Kornyshev AA, Kuznetsov AM, Ulstrup J, Walbran S. Mechanisms of proton conductance in polymer electrolyte membranes. *J Phys Chem B* 2001;105: 3646–62. https://doi.org/10.1021/jp003182s.

2 Nafion and modified Nafion membranes as PEMs

2.1 NAFION-THE COMMERCIAL MEMBRANE

Since the advancement of polymer electrolyte membrane-based fuel cells, perfluorosulfonic acid-based membranes are grabbing the market. Perfluorosulfonic acid-based membrane in the trade name of Nafion has been the popular one till date, and research is going on to find an alternative to this membrane at lower cost. Nafion, made by Dupont, is the commercial PEM material utilized without any exception to date. The polytetrafluoroethylene-based background of Nafion provides excellent thermal, mechanical, and chemical properties. The excellent proton conductivity of Nafion, in combination with thermal and chemical stability, makes it the most demanding material as PEM. Figure 2.1 displays the structure of some commercial Nafion manufactured by different companies to be utilized as PEM. Acidic side chains present in the Nafion are responsible for the proton conduction in the polymer. The nanophase-separated morphology of Nafion creates a long-interconnected proton conduction path to ease the passage of protons from one side of the electrode to the other. The hydrophilic sulfonic acid groups of the polymer transport protons in the presence of water molecules by forming hydronium ions. The standard terminology and specification of the Nafion 117 membrane are also summarized in Table 2.1.

Although, Nafion is utilized extensively in the field of PEMFCs, the demand for alternative membrane always exists due to high cost and loss of performance at the elevated temperature and/or progressive operation. Below, we will discuss the proton conduction mechanism in Nafion-based PEM in detail to understand the reason behind the loss of conductivity with operation. The proton conduction mechanism in Nafion suggests that the water serves a major role in deciding the proton conduction and proton movement in the membrane. With prolonged operation, the loss of water molecules with progressive drying of Nafion makes the conduction of proton difficult, and the proton conductivity is reduced with time. Below, we have discussed the water-assisted conduction mechanism in Nafion and perfluorosulfonic acid-based polymer electrolyte membranes.

DOI: 10.1201/9781003733744-2

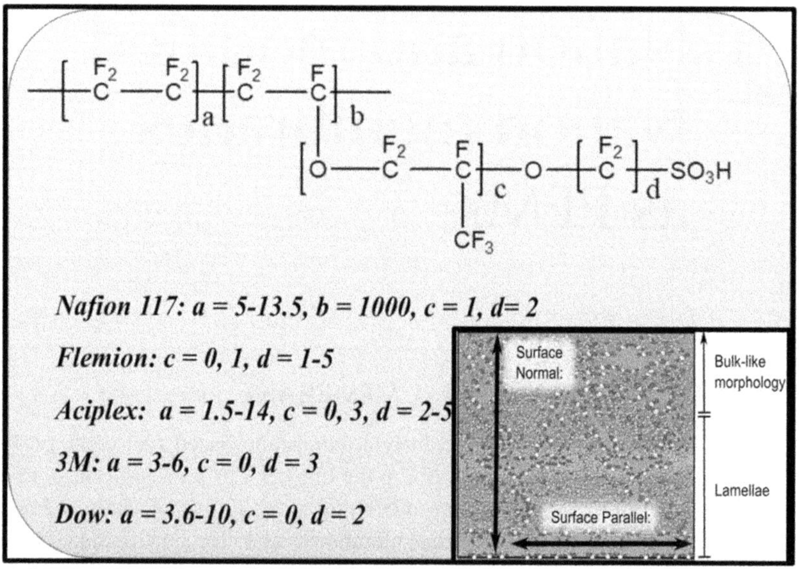

FIGURE 2.1 Structure and morphology of perfluorosulfonic acid PEM (Reprinted with permission from [1])

TABLE 2.1
Terminologies of Nafion 117, a standard PEM

Membrane terminology	Values	References
Proton conductivity, 25°C, hydrated	10^{-1} S cm^{-1}	[2]
Water uptake, soaked in water at 100°C for 1 hr.	38%	[2]
Ion exchange capacity	0.95–1.04 meqv g^{-1}	[2]
Tensile strength, 50% RH, 23°C	43 in MD and 32 in TD	[2, 3]
Methanol permeability	$2.1 \pm 0.2 \times 10^{-6}$ cm^2 s^{-1}	[4]

2.2 PROTON CONDUCTION IN NAFION

Proton exchange membranes have been extensively studied for fuel cells. The pure ionomers having negatively charged side chain functionalization, such as $-SO_3H$ and $-PO_4H_2$ are of prime interest due to the huge proton conductivity below 100°C. Nafion is the gold standard material in the field of polymer electrolytes. A suitable structural combination of hydrophilic and hydrophobic multiphase of Nafion contributes to its high proton conduction ability, chemical, and thermal stability. The polytetrafluoroethylene backbone of Nafion is responsible for high stability, however, the sulfonic acid functionalization favors the release of proton in water and facilitates proton conduction. In the case of Nafion, water plays a vital role in proton conduction since it dissociates the protons from the acidic groups and also eases the mobility of the hydrated protons. Therefore, the state of water plays a crucial role in efficient proton conduction. Since at low RH, with a decrement in water content, mobility of the hydrated protons gets hindered, and a huge reduction in proton conduction takes place.

2.3 HYDRATION LEVEL AND PROTON CONDUCTION IN PEM

The water sorption behavior of PEM is expressed in terms of hydration level, denoted by the symbol λ, defined as the number of water molecules sorbed per sulfonic acid group. The completely dehydrated condition, i.e. $\lambda = 0$, is not feasible since it requires the membrane to be heated up to a point, where the degradation of the membrane will start. About $\lambda = 1.5$ is considered to be remaining when the membrane is out of contact with any water [5]. The sorption of the very first water promotes the formation of the hydronium ion via the dissociation of sulfonic acid groups. The sulfonated head then acts as a nucleation site, and the remaining water forms a cluster around it. At the initial stage of hydration, the content of water is very low, and the insufficient amount of water is unable to provide a continuous path of water phase for proton conduction and hence the proton conductivity at this stage remains quite low. For $\lambda = [1, 2]$, the strength of the hydrogen bonds remains about 80% of that of pure water; however, the addition of more water to the cluster does not improve the strength a lot because of increasing cluster shape. When λ reaches 3–5, the cluster size continues to grow, and the excess protons remain mobile over the surface of the cluster. As soon as the λ value reaches greater than 2, some conduction pathway can be formed, allowing some proton conduction through the membrane. However, the membrane will still show low proton conductivity unless the λ value reaches 5, enabling the clusters to connect to each other and provide continuous conduction pathways. For Nafion, the primary hydration shell is made of $\lambda = [4, 6]$. At a still higher value of $\lambda \geq 6$, the clusters coalesce with each other and form larger clusters which lead to the formation of continuous conduction pathways. At this stage, the water mobility and water self-diffusion properties reach those of bulk water [5, 7]. The excess water or free water is shielded from the sulfonated head by the bound water molecules strongly connected in the primary hydration shells [7]. Figure 2.2 schematically represents the level of hydration with increasing λ value.

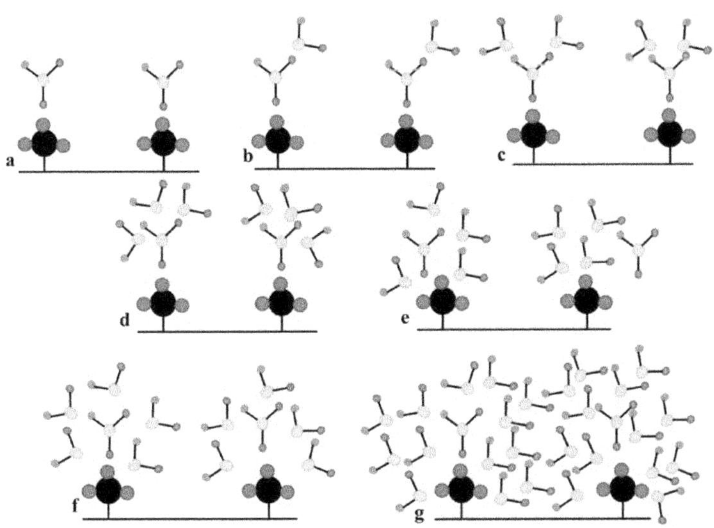

FIGURE 2.2 Hydration level of Nafion, (a) $\lambda = 1$, (b) $\lambda = 2$, (c) $\lambda = 3$, (d) $\lambda = 4$, (e) $\lambda = 5$, (f) $\lambda = 6$, and (g) $\lambda = 14$ (Concept taken from [8])

Phase separation of Nafion in micro-state of Nafion is discussed [9]. Figure 2.3 represents pictorially the hydrophobic perfluorosulfonic acid backbone of Nafion and clusters of sulfonic acid hydrophilic groups that absorb water inside the pores. The study revealed that in the absence of water molecules, the dipoles present in the sulfonic acid will try to form hydrophilic channel. With progressive absorption of water inside, the hydrophilic channel will push the hydrophobic matrix and create free volume to allow more and more water molecules to penetrate inside. The development of the first hydration shell allows further to form the second hydration shell by the addition of more water molecules inside the free space of the hydrophilic channel.

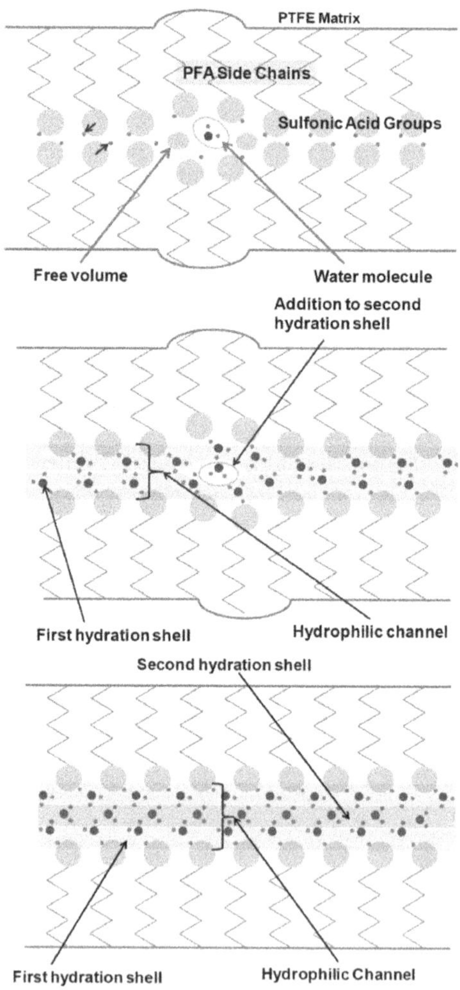

FIGURE 2.3 Pictorial representation of multilayer water adsorption (Reprinted with permission from [9])

Subsequent addition of water beyond the second shell shifts the interaction from adsorption to free space to more toward bulk water. The adsorption of water is now similar to multilayer adsorption.

However, the proton conductivity of these membranes declines with the progressive drying of the membranes since proton conduction is purely a water-dependent process for Nafion. High cost and loss of property during progressive operation open up a new window for the development of membranes other than Nafion.

2.4 CHALLENGES ASSOCIATED WITH NAFION

To overcome these shortcomings of Nafion, research and development of an alternative PEM is the primary concern of scientific society working in this field. Plenty of research groups are involved worldwide in the research and development of PEMFC. In this chapter, we will discuss the various classes of polymer matrices utilized as PEM, their prospective as PEM, and future trend of research

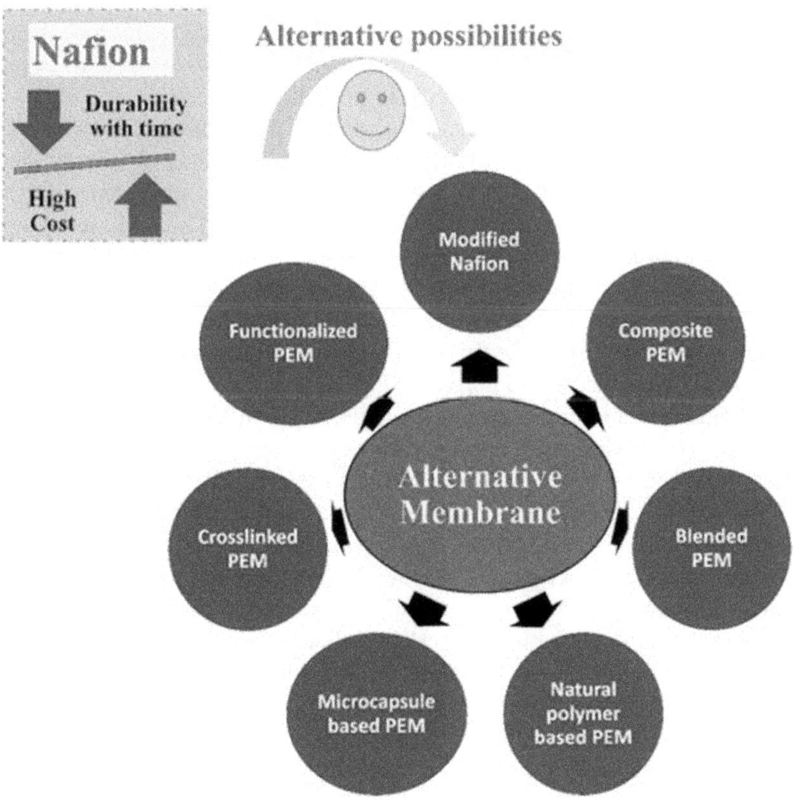

FIGURE 2.4 Membranes alternative to Nafion

in detail (Figure 2.4). The mechanism of proton conduction through the membrane is related to the water content of the membrane which is still a critical issue. The conduction mechanism of PEM with respect to the level of hydration has been explained to understand the conduction process. In addition, this book also focuses on the effect of different factors controlling the performance of the device and the approaches that are to be taken to manage water and heat inside a PEMFC in a better way.

2.5 SCOPE OF MODIFIED NAFION MEMBRANE

The proton conduction property of perfluorosulfonic acid membranes is greatly affected by the water content of the membranes. The main contributing factor affecting the performance of a PEM is the amount of bound water in the membrane, i.e., the amount of water tightly bounded to the hydrophilic sulfonic acid groups of a polymer matrix that do not evaporate at 100°C under the relatively anhydrous condition of the membrane. The bound water inside the membrane contributes to the conduction mechanism via the Grotthuss pathway through hydrogen bond formation among the water molecules under reduced humidity. Hence, an enhancement in the bound water content of the membrane is an effective approach to improve the proton conduction ability of Nafion membrane to be operated under low relative humidity. The morphological difference in Nafion and SPEEK membranes was observed by TEM analysis [10, 11]. Devi et al. have investigated the content of free, bound, and bulk water for hydrated Nafion and SPEEK membranes using Terahertz spectroscopy, where a continuous decline in water content has been observed with time as evidenced from molecular quantification in Figure 2.5 [12]. The study reveals that % water content in SPEEK under hydration at the initial stage is about 2.36 times more compared to Nafion due to the highly porous structure of SPEEK. Again, a better water retention capacity specifically more bound water content in SPEEK at a longer duration is attributed to the presence of strong hydrogen bonding between -SO_3H and H_2O in widely distributed small pores of SPEEK.

Nafion has been modified by various approaches in earlier studies such as by treating with phosphoric acid, doping with cesium ion impregnated with polymer precursor followed by in situ polymerization approach [13–15]. The surface structure of Nafion 117 has been modified using a low dose electron beam (EB) to improve the performance of PEM for DMFC application [16]. A small dose of radiation is effective to reduce the methanol permeation of the membrane with an additional advantage of improved power output. Nafion is obtained by co-polymerization of tetrafluoroethylene and unsaturated perfluoroalkyl sulfonyl fluoride, where the latter is used in variable content to define the polymer structure. Nafion of varying equivalent weights is available, such as 900, 110, and 1200. The equivalent weight of a Nafion is defined as a gram of dry Nafion per mole of hydrophilic sulfonic acids provided the material remains in its acidic form. The study claims that both the diffusion of water and proton near the fluorocarbon and acidic groups lining the hydrated channel are much faster compared to that within the channels of water as reflected in diffusion constants [10].

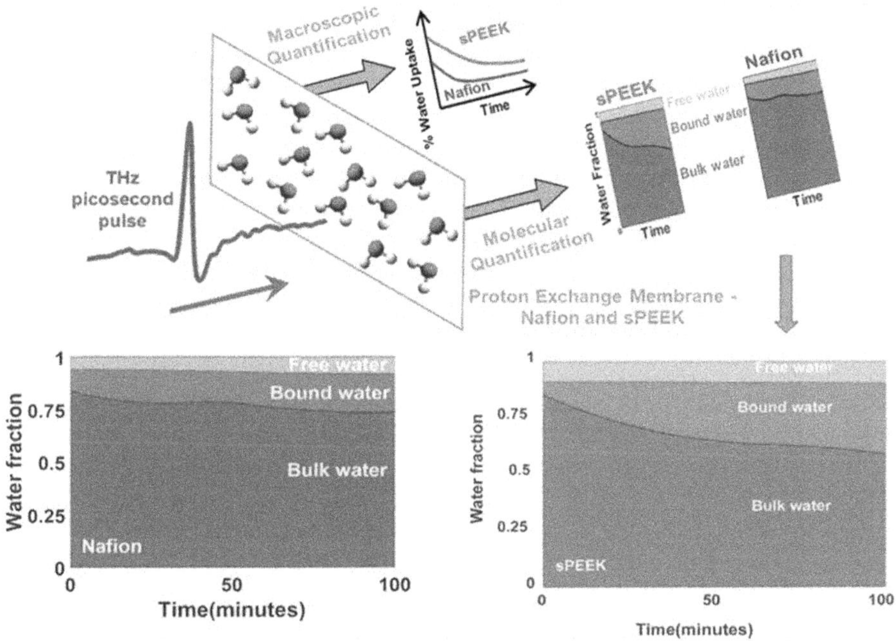

FIGURE 2.5 A relative fraction of water in Nafion and SPEEK membranes with time determined by Terahertz spectroscopy. Reprinted with permission from [12]

2.6 RECENT DEVELOPMENT IN NAFION-BASED COMPOSITE MEMBRANES

The proton conductivity of Nafion depends on the equivalent weight, pre-treatment process of membranes, and equilibrium parameters. The proton conduction in Nafion is a water-assisted process, where water plays a dominant role for the ease of movement of the hydrated protons. The high proton conductivity of Nafion drops significantly with a reduction in water content since the bridging between the ionic clusters is lost in the process. To address the reduction in proton conductivity with progressive drying of the PEM during prolonged operation, water retention inside the membrane is improved by addition of hydrophilic fillers and/or introduction of second polymer to form the blend with Nafion where the second polymer may or may not have sulfonic acid or similar hydrophilic groups as functionality. A common practice is to incorporate hydrophilic inorganic fillers such as silica, titania, or zirconia oxide and zeolites to maintain water at high temperature and low humidity. The addition of these hydrophilic fillers improves the binding energy of water with the additional benefit of enhancement in acidic sites both in numbers and strengths. The inorganic fillers narrow down the size of hydrophilic channels from 7.9 to 6.5 nm contributing to a significant improvement in proton conduction. Again, the affinity of filler particles to absorb water helps in the retention of number of molecules of water within the membrane. To enhance the proton conductivity researchers have also introduced side chain functionality in the inorganic fillers to make sulfonated, phosphonated, or carboxylated metal oxides. The additional hydrophilic groups in the nanofiller also contribute to the overall improvement in the proton conductivity of the polymer nanocomposite-based PEMs. In addition to inorganic oxides, solid proton conductors, such as heteropolyacids and phosphates (Ce, Ti, Zr), have also been explored to examine their effect on proton transport property [3]. The proton conducting fillers, ionic liquids, etc., provide additional protons availability in the membranes and enhance the overall performance of the device.

To address the inherent problem of perfluorosulfonic acid based membranes, research is going on as discussed above. Sigwadi et al. have modified Nafion membranes with water-retentive mesoporous zirconia nanoparticles [17]. Nafion has also been modified by heteropolyacids doped silica specifically with phosphotungstic and silicotungstic acids doped silica for high-temperature DMFC devices [18, 19]. Nafion/silica composite membranes are investigated by Sahu et al. The silica particles synthesized by common sol-gel techniques are embedded in the PFSA membrane via water hydrolysis [20]. The mechanical and electrical properties of Nafion and titania-filled Nafion composite membranes are studied under controlled conditions [21]. The addition of titania enables low-strain hardening in composite membranes. Chen et al. have developed a nanocomposite membrane based on Nafion and acidic zeolite beta for DMFC devices [22]. The nanocomposite membranes are characterized by a remarkable reduction in methanol permeability of 40% with slight reduction in proton conductivity of these membranes.

In an interesting study, Ketpang et al. have developed Nafion composite membrane containing various mesoporous metal oxide nanotubes, such as titania, ceria, and zirconia, for fuel cells to be operated under low RH [23]. The tubular morphology helps to produce 1.29 times better current density compared to commercial Nafion. The incorporation of mesoporous fillers in tubular morphology is beneficial due to the effective water diffusion through the membrane. This is because of the electro-osmotic and back diffusion process. The combined effect of water retention and diffusion ability effectively suppresses the Ohmic resistance of the membrane with improved catalyst utilization at the cathode side of the MEA. Ketpang et al. have developed a nanocomposite membrane using Nafion as the matrix and porous zirconium oxide nanotubes (ZrNT) as the nanofiller for low temperature and low humidity fuel cell operating conditions [24]. The water retention properties of the porous ZrNTs are expected to improve the performance of fuel cells under reduced humidity via favorable water diffusion through the perfluorosulfonic acid membranes.

Segale et al. investigated the effect of TMS coated silica nanoparticles (synthesized from sol gel approach) on the fuel cell properties of Nafion 117 [25]. The size of monodispersed nanoparticles was in the range of 150–350 nm. The nanocomposite membranes were characterized by various properties such as thermal and mechanical stability, methanol permeability, proton conduction, and ion exchange capacity. The nanocomposite membranes at 5 wt.% filler loading exhibited good enhancement in water uptake and proton conductivity. The incorporation of nanofiller increased the proton conduction along with water retention capacity which confirmed its candidature in low to medium temperature fuel cell applications. In another study, Nafion composite membranes were fabricated by adding sulfo ethyl cellulose at weight percentage range of 5–10 wt.%. The presence of sulfo ethyl cellulose affected the hydrophilic nature of the membranes with improvement in water absorption capacity. The proton conductivity was reported to be as high as 170 mS cm^{-1} at a temperature of around 120°C. The structural investigation indicated that crosslinking between the sulfonated groups of Nafion and hydroxyl groups of sulfo ethyl cellulose exist. The results revealed that the interaction between the groups leads to reduced size of crystallites and increased peak broadening in X-ray diffraction analysis. The composite membranes are also investigated for morphological and thermal analysis to establish its potential in polymer electrolyte membranes [26]. In another study, the significance of nanofillers on the operation and performance of Nafion-based membranes were investigated. Different types of nanofillers, such as inorganic oxides, carbon nanofillers, metal organic frameworks, and polyoxometalate, were discussed [27]. The synthesis method, nanofiller type, nanofiller morphology, and structure property relationship between the matrix and the nanomaterials contribute to the overall performance of the nanocomposite membranes. Nafion has also been blended with other polymers such as PVDF to fabricate films by uniaxial stretching approach [28]. The Nafion with phase separated morphology forms proton channels for increased conductivity due to presence of hydrophilic domains whereas the PVDF by uniaxial stretching exhibited highly aligned beta crystalline phases. The blended membranes contributed to the enhancement in proton conduction ability of Nafion leading to synergism in performance. The structural interaction between the polymeric phases was examined using small angle X-ray scattering and the atomic force microscopy. The proton conduction has been reported to improve when measured in-plane due to increased conductance at the membrane catalyst interfacial area which eventually contributed to higher power generation in the device.

Addition of ionic liquid in the Nafion matrix remains another strategy to improve the inherent proton conductivity of Nafion specifically for high-temperature fuel cells. Ionic liquids are organic salts at room temperature having melting point much below 100°C. Ionic liquids contain organic and inorganic cations and anions. The incorporation of ionic liquid enhances the overall proton conductivity of Nafion due to the availability of more active protons. The imidazole-containing ionic liquids lead to the formation of hydrogen-bonded structure facilitating proton hopping through the membrane in lower water content also. The low vapor pressure of ionic liquids also favors the retention of this proton conductor at high temperature, making them a suitable material for the utilization in high temperature and low humidity conditions. Yang et al. have incorporated BMIm into Nafion followed by doping with phosphoric acid [29]. The composite membrane so obtained exhibited lower methanol crossover compared to Nafion 115 with improved proton conductivity at high temperature and low humidity. The conduction of proton is governed by proton hopping mechanism due to the formation of hydrogen bond with phosphoric acid. The doping of phosphoric acid also contributes to retaining the required thermal and mechanical stability. The addition of ionic liquid remains a suitable option for the fabrication of modified Nafion membranes; however, proper choice of ionic liquid and extent of loading should be optimized to retain the durability and mechanical integrity of the membranes along with leakage of low melting proton conduction during the long-term operation.

Carbon nanotubes, graphene, and fullerene are often added in the Nafion matrix for PEM to be utilized in DMFC applications [30]. The nanocomposites prepared using carbon nanomaterials sound promising for DMFC due to excellent mechanical properties and resistance to methanol permeation. However, addition of these fillers remains challenging due to difficulty in uniform dispersion of these fillers inside the polymer matrix restricting improvement in overall performance of the PEM. The poor dispersion of carbon nanomaterials arises as a result of extensive aggregation leading to difficulty in proton transport and overall reduction in mechanical strength. Again with loading, the density of sulfonic acid functional content decreases in Nafion as well. To improve the dispersion of the filler inside the Nafion, carbon is either functionalized or grafted with functional groups such as chitosan and silica thereby improving the interfacial interaction of fillers with the polymer, leading to better distribution [31]. Sulfonated CNT/Nafion composite membrane exhibited improved proton conductivity compared to the pure Nafion and Nafion/CNT composite as per literature [32]. The introduction of hydrophilic functional groups in CNT provides additional source of protons to the membrane and helps to grow ion conduction channels due to more bound water. The sulfonations of CNT under oxidizing condition generate oxidative defects and thereby raise the electrical resistance whereas the presence of sulfonate functional groups provides additional protons. The combination of two leads to comparable proton conductivity as that of Nafion with added benefit of improved mechanical and thermal stability for long-term operation [33]. Carbon dots are zero-dimensional carbon nanomaterial of diameter of about 10 nm or less. Carbon dots are added in Nafion to form a composite containing carboxylic and sulfonic acid moieties [34]. The Nafion composite membranes suitable for high-temperature application contain multiple types of functionalities introduced by hydrogen bonding. The composite membrane containing carbon nanodot displays five times higher proton conductivity and halved methanol permeation compared to pristine Nafion membrane indicating promising alternative PEM.

Graphene, a two-dimensional single-atomic-layered carbon nanomaterial, is an interesting material class owing to its large number of functional groups and thermo-mechanical stability. A single-layered graphene has the capability to restrict permeability of all atoms except proton. Graphene has surface functional groups, such as carboxyl and hydroxyl, that provides hydrophilicity. The oxygenated sites available give provision to introduce further functionalization by adding more proton source to the system [35]. Graphene oxide also has the ability to restrict the flow of the methanol inside the membrane making them a suitable choice for DMFC devices [36, 37].

2.6.1 Structural Modification of Nafion

The physical modification of the membrane is another approach that has attracted the attention of the researchers in order to develop PEM without modifying its bulk properties. Lee et al. have worked on the modification of Nafion by irradiation and ion implantation [38]. The irradiation introduces protons in the surface of Nafion by breaking the chemical bonds. The high energy bombardment leads to chain scission in the hydrophilic side chains generating active radicals. An interaction of these radicals with atmospheric oxygen forms polar functional groups, such as carboxylic, hydroxyl, and carbonyl, that contribute to the enhanced proton conductivity in Nafion due to efficient water retention inside the membrane. The membranes are reported to exhibit 58% reduced methanol permeation along with desired proton conductivity. In another approach, Nafion membranes are processed by two different processing methods and examined for their proton conductivity [39]. One membrane is processed by extrusion and other by solution casting method. The extruded membranes are reported to be of higher conductivity than the solution casted membranes since the size of water channels in extruded Nafion is smaller as compared to solution-casted one. This in turn helps Nafion to hold higher amount of non-bulk water, i.e., the water molecules in close proximity of sulfonic acid groups. Hence, the extruded membranes exhibit high conductivity due to easier and rapid transportation of protons. Researchers also tried to observe the effect of size of ionic clusters of Nafion on the proton conductivity [40]. The ionic cluster size is varied by UV treatment, and effect of cluster size on the methanol permeation is observed. The UV exposure improves the ionic interaction between the acidic groups creating favorable hydrophilic channels that facilitate proton transfer. The power density is reported to be improved by 1.2 to 1.5 times under optimum UV exposure [41].

2.7 COMMERCIAL MEMBRANES AND ITS UTILITY

A number of commercial membranes of perfluorosulfonic acid class have been utilized as commercial polymer electrolyte membranes. The commercial perfluorosulfonic acid-based membranes mainly differ in thickness, basis weight, or equivalent weight and thereby the properties of interest. As per the nomenclature, the commercial Nafion membranes are described based on their equivalent weight that represents the gram of dry Nafion per mole of sulfonic acid. The 117 number indicates that 110 gram of dry Nafion is present per mole of sulfonic acid groups in its acidic form and the membrane should have a thickness of 0.007 inches by definition. The thickness (μm) and basis weight (g m^{-2}) of Nafion membranes of commercial importance are enlisted in Table 2.2 along with level of hydration, ion exchange capacity and water uptake to understand the importance of these commercial membranes.

2.8 CHAPTER SUMMARY

This chapter discusses about the perfluorosulfonic acid-based membranes, their commercial importance, proton conduction mechanism, challenges, and possible remedies. In this chapter, various modes of proton conduction in Nafion have been discussed in detail in order to understand the role of water in overall conduction process. To address the issues related to the Nafion, various approaches have been adopted in the literature, and modification of Nafion membranes is still going on to obtain suitable alternative membrane. The scope of Nafion-based composites and nanocomposites with various fillers, such as inorganic oxides, heteropolyacids, ionic liquids, and water retentive fillers, studied by various groups, have been

TABLE 2.2
Specification of perfluorosulfonic acid membranes

Membrane	Thickness (um)	Basis weight (g m^{-2})	Ion exchange capacity (mmol. g^{-1})	Water uptake (%)	Hydration (H_2O/ SO_3H)	References
N115	127	250	1.04 ± 0.02	39.9 ± 2.9	21.2 ± 1.2	[3, 42, 43]
N117	183	360	1.04 ± 0.00	39.9 ± 0.1	21.4 ± 0.0	[3, 42, 43]
NR211	25.4	50	1.22 ± 0.01	28.1 ± 13.6	12.7 ± 6.1	[3, 42, 43]
NR212	50.8	100	1.16 ± 0.01	38.6 ± 6.8	18.6 ± 3.3	[3, 42, 43]
NR111	25.4	50	-	-	-	[3, 43]
NR112	50.8	100	0.98 ± 0.00	-	20.7 ± 0.5	[44, 45]
N105	127	-	1.10 ± 0.00	47.9 ± 0.3	24.1 ± 0.1	[3, 42, 43]
N1035	88.5	175	1.15 ± 0.00	46.1 ± 3.8	22.3 ± 1.9	[3, 42, 43]
N1135	89	190	1.05 ± 0.02	39.3 ± 2.4	20.8 ± 1.0	[3, 42, 43]
N1110	254	500	-	-	-	[43]

included. The chapter aims to provide the discussion on the possible ideas that have been tried by the researchers to address the reduction of proton conductivity with time in perfluorosulfonic acid-based membranes such that highly conductive Nafion membranes can be utilized over longer period without loss of the performance. Other membranes as an alternative to Nafion have been discussed in succeeding chapters of this book.

REFERENCES

[1] DeCaluwe SC, Baker AM, Bhargava P, Fischer JE, Dura JA (2018) Structure-property relationships at Nafion thin-film interfaces: Thickness effects on hydration and anisotropic ion transport. *Nano Energy* 46:91–100. https://doi.org/10.1016/j.nanoen.2018.01.008.

[2] www.fuelcellstore.com.

[3] Sahu AK, Pitchumani S, Sridhar P, Shukla AK (2009) Nafion and modified-Nafion membranes for polymer electrolyte fuel cells: An overview. *Bull Mater Sci* 32:285–294. https://doi.org/10.1007/s12034-009-0042-8.

[4] Li X, Zhao Y, Feng Z, Xiang X, Wang S, Xie X, Ramani VK (2017) Ring-opening metathesis polymerization for the preparation of polynorbornene-based proton exchange membranes with high proton conductivity. *J Memb Sci* 528:55–63. https://doi.org/10.1016/j.memsci.2016.12.050.

[5] Laporta M, Pegoraro M, Zanderighi L (1999) Perfluorosulfonated membrane (Nafion): FT-IR study of the state of water with increasing humidity. *Phys Chem Chem Phys* 1:4619–4628. https://doi.org/10.1039/a904460d.

[6] Choi P, Datta R (2003) Sorption in proton-exchange membranes. *J Electrochem Soc* 150:E601–E608. https://doi.org/10.1149/1.1623495.

[7] Kreuer KD (2001) On the development of proton conducting polymer membranes for hydrogen and methanol fuel cells. *J Memb Sci* 185:29–39. https://doi.org/10.1016/S0376-7388(00)00632-3.

[8] Fimrite J, Struchtrup H, Djilali N (2005) Transport phenomena in polymer electrolyte membranes. *J Electrochem Soc* 152:A1804–A1814. https://doi.org/10.1149/1.1952627.

[9] Zhao Q, Majsztrik P, Benziger J (2011) Diffusion and interfacial transport of water in Nafion. *J Phys Chem B* 115:2717–2727. https://doi.org/10.1021/jp1112125.
[10] Song J, Han OH, Han S (2015) Nanometer-scale water- and proton-diffusion heterogeneities across water channels in polymer electrolyte membranes. *Angew Chem Int Ed* 54:3615–3620. https://doi.org/10.1002/anie.201408318.
[11] Wang R, Liu S, Wang L, Ming L, Gao C (2019) Understanding of nanophase separation and hydrophilic morphology in Nafion and SPEEK membranes: A combined experimental and theoretical studies. *Nanomaterials* 9:869–881. https://doi.org/10.3390/nano9060869.
[12] Devi N, Ray S, Shukla A, Bhat SD, Pesala B (2019) Non-invasive macroscopic and molecular quantification of water in Nafion® and SPEEK Proton Exchange Membranes using terahertz spectroscopy. *J Memb Sci* 588:117183–117193. https://doi.org/10.1016/j.memsci.2019.117183.
[13] Tricoli V (1998) Proton and methanol transport in poly(perfluorosulfonate) membranes containing Cs + and H + cations. *J Electrochem Soc* 145:3798–3801. https://doi.org/10.1149/1.1838876.
[14] Wainright JS, Wang J - T., Weng D, Savinell RF, Litt M (1995) Acid-doped polybenzimidazoles: A new polymer electrolyte. *J Electrochem Soc* 142:L121–L123. https://doi.org/10.1149/1.2044337.
[15] Jia N (1999) Modification of Nafion proton exchange membranes to reduce methanol crossover in PEM fuel cells. *Electrochem Solid-State Lett* 3:529–531. https://doi.org/10.1149/1.1391199.
[16] Hobson LJ, Ozu H, Yamaguchi M, Hayase S (2001) Modified Nafion 117 as an improved polymer electrolyte membrane for direct methanol fuel cells. *J Electrochem Soc* 148:A1185–A1190. https://doi.org/10.1149/1.1402980.
[17] Sigwadi R, Dhlamini M, Mokrani T, Ṋemavhola F (2017) Wettability and mechanical strength of modified nafion® nanocomposite membrane for fuel cell. *Dig J Nanomater Biostruct* 12:1137–1148
[18] Ramani V, Kunz HR, Fenton JM (2005) Stabilized heteropolyacid/Nafion® composite membranes for elevated temperature/low relative humidity PEFC operation. *Electrochim Acta* 50:1181–1187. https://doi.org/10.1016/j.electacta.2004.08.015.
[19] Staiti P, Aricò AS, Baglio V, Lufrano F, Passalacqua E, Antonucci V (2001) Hybrid Nafion–silica membranes doped with heteropolyacids for application in direct methanol fuel cells. *Solid State Ion* 145:101–107. https://doi.org/10.1016/S0167-2738(01)00919-5.
[20] Sahu AK, Selvarani G, Pitchumani S, Sridhar P, Shukla AK (2007) A sol-Gel modified alternative Nafion-silica composite membrane for polymer electrolyte fuel cells. *J Electrochem Soc* 154:B123–B132. https://doi.org/10.1149/1.2401031.
[21] Satterfield MB, Majsztrik PW, Ota H, Benziger JB, Bocarsly AB (2006) Mechanical properties of Nafion and titania/Nafion composite membranes for polymer electrolyte membrane fuel cells. *J Polym Sci B Polym Phys* 44:2327–2345. https://doi.org/10.1002/polb.20857.
[22] Chen Z, Holmberg B, Li W, Li W, Wang X, Deng W, Munoz R, Yan Y (2006) Nafion/Zeolite nanocomposite membrane by in situ crystallization for a direct methanol fuel cell. *Chem Mater* 18:5669–5675. https://doi.org/10.1021/cm060841q.
[23] Ketpang K, Lee K, Shanmugam S (2014) Facile synthesis of porous metal oxide nanotubes and modified Nafion composite membranes for polymer electrolyte fuel cells operated under low relative humidity. *ACS Appl Mater Interfaces* 6:16734–16744. https://doi.org/10.1021/am503789d.
[24] Ketpang K, Son B, Lee D, Shanmugam S (2015) Porous zirconium oxide nanotube modified Nafion composite membrane for polymer electrolyte membrane fuel cells operated under dry conditions. *J Memb Sci* 488:154–165. https://doi.org/10.1016/j.memsci.2015.03.096.

[25] Segale ME, Mokrani T, Sigwadi RA (2024) Synthesis and characterization of enhanced proton-conducting Nafion®117- silica composite membranes for fuel cell applications. *J Nano Res* 82:95–116. https://doi.org/10.4028/p-3lgu0L.

[26] Charradi K, Landolsi Z, Heinze T, Brahmia B, Chtourou R, Keshk S (2024) Sulfo ethyl cellulose/ Nafion composite for high-temperature proton exchange membrane. *J Appl Polym Sci* 141. https://doi.org/10.1002/app.55665.

[27] Song S, He H, Chai S, Li H (2024) Advanced Nafion/nanofiller composite proton exchange membranes for fuel cell applications. *Polymer (Guildf)* 307:127241. https://doi.org/10.1016/j.polymer.2024.127241.

[28] Lee I, Lee J, Lee D, Kim S, Kim SK, Bae I (2024) Synergistic molecular alignment and dipole polarization in stretched Nafion/Poly(vinylidene fluoride) blend membranes for high proton conduction in PEMFCs. *ACS Appl Mater Interfaces* 16:42164–42175. https://doi.org/10.1021/acsami.4c06637.

[29] Yang J, Che Q, Zhou L, He R, Savinell RF (2011) Studies of a high temperature proton exchange membrane based on incorporating an ionic liquid cation 1-butyl-3-methylimidazolium into a Nafion matrix. *Electrochim Acta* 56:5940–5946. https://doi.org/10.1016/j.electacta.2011.04.112.

[30] Zhang D, Xin L, Xia Y, Dai L, Qu K, Huang K, Fan Y, Xu Z (2021) Advanced Nafion hybrid membranes with fast proton transport channels toward high-performance vanadium redox flow battery. *J Memb Sci* 624:119047–119056. https://doi.org/10.1016/j.memsci.2020.119047.

[31] Rambabu G, D. Bhat S, Figueiredo FML (2019) Carbon nanocomposite membrane electrolytes for direct methanol fuel cells—A Concise review. *Nanomaterials* 9:1292–1321. https://doi.org/10.3390/nano9091292.

[32] Yin C, Li J, Zhou Y, Zhang H, Fang P, He C (2018) Enhancement in proton conductivity and thermal stability in nafion membranes induced by incorporation of sulfonated carbon n1anotubes. *ACS Appl Mater Interfaces* 10:14026–14035. https://doi.org/10.1021/acsami.8b01513.

[33] Asgari MS, Nikazar M, Molla-abbasi P, Hasani-Sadrabadi MM (2013) Nafion®/histidine functionalized carbon nanotube: High-performance fuel cell membranes. *Int J Hydrogen Energy* 38:5894–5902. https://doi.org/10.1016/j.ijhydene.2013.03.010.

[34] Jia W, Tang B, Wu P (2018) Carbon dots with multi-functional groups and the application in proton exchange membranes. *Electrochim Acta* 260:92–100. https://doi.org/10.1016/j.electacta.2017.11.047.

[35] Sha Wang L, Nan Lai A, Xiao Lin C, Zhang QG, Zhu AM, Liu QL (2015) Orderly sandwich-shaped graphene oxide/Nafion composite membranes for direct methanol fuel cells. *J Memb Sci* 492:58–66. https://doi.org/10.1016/j.memsci.2015.05.049.

[36] Parthiban V, Akula S, Peera SG, Islam N, Sahu K (2016) Proton conducting nafion-sulfonated graphene hybrid membranes for direct methanol fuel cells with reduced methanol crossover. *Energy & Fuels* 30:725–734. https://doi.org/10.1021/acs.energyfuels.5b02194.

[37] Yuan T, Pu L, Huang Q, Zhang H, Li X, Yang H (2014) An effective methanol-blocking membrane modified with graphene oxide nanosheets for passive direct methanol fuel cells. *Electrochim Acta* 117:393–397. https://doi.org/10.1016/j.electacta.2013.11.063.

[38] Lee J-S, Hwang I-T, Jung C-H, Choi J-H (2016) Surface modification of Nafion membranes by ion implantation to reduce methanol crossover in direct methanol fuel cells. *RSC Adv* 6:62467–62470. https://doi.org/10.1039/C6RA12756H.

[39] Ling X, Bonn M, Parekh SH, Domke KF (2016) Nanoscale distribution of sulfonic acid groups determines structure and binding of water in Nafion membranes. *Angew Chem Int Ed* 55:4011–4015. https://doi.org/10.1002/anie.201600219.

[40] Li L, Su L, Zhang Y (2012) Enhanced performance of supercritical CO2 treated Nafion 212 membranes for direct methanol fuel cells. *Int J Hydrogen Energy* 37:4439–4447. https://doi.org/10.1016/j.ijhydene.2011.11.110.

[41] Rao AS, Rashmi KR, Manjunatha DV, Jayarama A, Prabhu S, Pinto R (2019) Pore size tuning of Nafion membranes by UV irradiation for enhanced proton conductivity for fuel cell applications. *Int J Hydrogen Energy* 44:23762–23774. https://doi.org/10.1016/j.ijhydene.2019.07.084.

[42] Albert A, Barnett AO, Thomassen MS, Schmidt TJ, Gubler L (2015) Radiation-grafted polymer electrolyte membranes for water electrolysis cells: Evaluation of key membrane properties. *ACS Appl Mater Interfaces* 7:22203–22212. https://doi.org/10.1021/acsami.5b04618.

[43] Yang Y, Siu A, Peckham TJ, Holdcroft S. (2008) Structural and morphological features of acid-bearing polymers for PEM fuel cells. Editor: Scherer, GG. *Fuel Cells I. Advances in Polymer Science.* Springer Berlin Heidelberg, Berlin, Heidelberg, p. 55–126, 2008. https://doi.org/10.1007/12_2008_134

[44] Slade SM, Ralph TR, Ponce de León C, Campbell SA, Walsh FC (2010) The ionic conductivity of a nafion® 1100 series of proton-exchange membranes re-cast from butan-1-ol and propan-2-ol. *Fuel Cells* 10:567–574. https://doi.org/10.1002/fuce.200900118.

[45] Slade S, Campbell SA, Ralph TR, Walsh FC (2002) Ionic conductivity of an extruded nafion 1100 ew series of membranes. *J Electrochem Soc* 149:A1556–A1564. https://doi.org/10.1149/1.1517281.

3 Functionalized polymer electrolyte membranes

3.1 FUNCTIONALIZED PEM FOR LOW-TEMPERATURE FUEL CELLS

The microstructure of PEM plays an important role in determining proton conduction ability of the membrane. The standard Nafion® membrane possesses a long-interconnected channel of hydrophilic sulfonic acid groups, which enables them to conduct protons efficiently. The presence of phase-separated hydrophilic-hydrophobic domains plays a crucial role in determining the conduction behavior [1]. In the search for alternative PEM that can be a suitable substitute to costly perfluorosulfonic acid membranes, functionalization approach is adopted. The introduction of hydrophilic functionality in the polymer backbone creates a combination of hydrophilic and hydrophobic domains similar to commercial Nafion®. The extent and type of functionalization are again a matter of concern to regulate the proton conduction in PEM. To enhance the availability of hydrophilic moieties inside the PEM various types of functionalization have been introduced in to the base polymers, such as sulfonation, phosphonation, florination, and carboxylation. Sulfonation remains the easiest and most common approach and has been explored extensively. In the following section, we will discuss the effect of different types of functionalization on different properties of base polymers and their ultimate effect on the overall performance of PEMFC devices.

3.2 SULFONATED PEM

Sulfonation is a widely accepted method to impart phase separation in polymer matrices suitable for fuel cell membranes due to the insertion of acidic hydrophilic sulfonic acid groups similar to standard commercial Nafion® membranes. However, the proton conductivity of these membranes is dependent on the extent of sulfonation. A very high degree of sulfonation provides good proton conduction due to the large numbers of acidic groups; however, the dimensional stability of these highly sulfonated membranes is reduced which have a detrimental effect on the mechanical integrity of the membrane. Hence, an optimum degree of sulfonation is needed to perform as an efficient PEM [2]. Due to the extensive convenience, sulfonation has been employed in various polymer matrices, such as polyether ether ketone (PEEK), polysulfones (PSU), polyimides (PI), polyphenylenes (PPS), polyphosphazenes, and polyarylenes. The sulfonation of PEEK is extensively studied by various researchers. Sulfonated polyether ether ketone (SPEEK) is combined with various inorganic fillers and polymers to prepare polymer nanocomposite membranes [3–5]. Figure 3.1 represents a schematic of the transport of water by back diffusion from cathode to

DOI: 10.1201/9781003733744-3

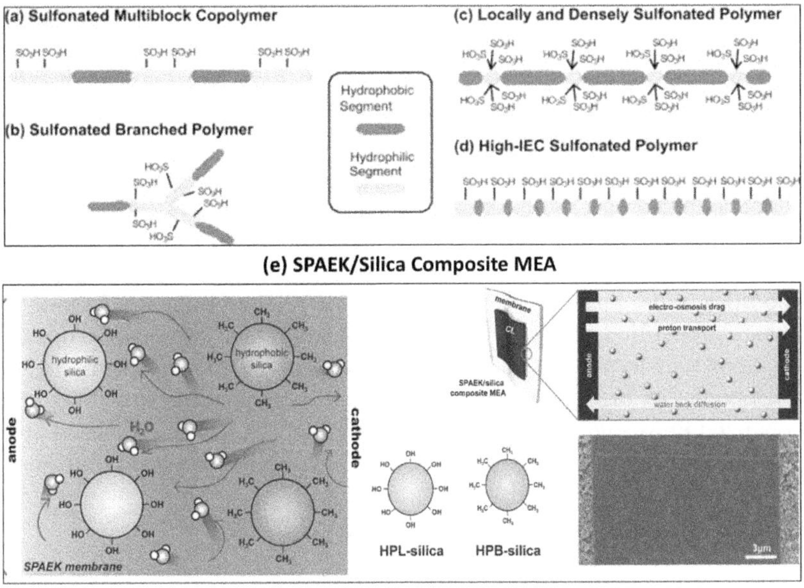

FIGURE 3.1 (a–d) Different architectures possible in sulfonated PEM [7], (e) representation of movement of water in bilayer asymmetric composite PEM and mechanism of conduction by back diffusion [6] (Reprinted with permission)

anode side in silica-modified sulfonated poly(arylene ether ketone) type asymmetric nanocomposite membrane [6]. The movement of water inside the cell during operation has been illustrated. An electro-osmotic drag has been evidenced from anode to cathode side due to the cell potential. The transport of protons is mainly governed by two possible mechanisms as discussed earlier. The water generated during fuel cell operation creates a gradient that serves as the major contributing factor for the back-diffusion process inside the membrane as evidenced in Figure 3.1.

The choice of sulfonating agent for post-functionalization of polymer matrix remains an essential criterion due to the solubility and degradation issues. Strong sulfonating agents, e.g., conc. H_2SO_4, chlorosulfonic acid, and fuming sulfuric acid, and mild sulfonating agents, e.g., acetyl sulfate and sulfur trioxides, are commonly used. SPEEK/PBI ionically crosslinked nanocomposite PEM has been studied recently [8]. The ionic crosslinks help to improve the mechanical strength and oxidative stability of these membranes with reduced swelling and methanol permeation. PBI-based composite membranes have also been prepared with sulfonated polyvinyl alcohol (sPVA). The crosslinked composite membrane shows good methanol permeability of 10^{-8} cm^2 s^{-1} with proton conductivity in the range of 10^{-3} S cm^{-1} [9]. In another work, Kim et al. have demonstrated sulfonated poly(phenylene)s both branched and linear polymers for fuel cells [10]. The membrane prepared of branched polymer exhibits proton conductivity of 94.6 mS cm^{-1} at 80°C and 90% RH and better dimensional stability that is comparable to Nafion 211. The membrane made of branched polymer shows a power density of 0.60 W cm^{-2} which is comparable to standard Nafion 212. The effect of degree of sulfonation on polyethersulfone membranes shows that sulfonation of 80 mol% sulfonic acid monomer in the polymer main chain leads to the proton conductivity of 62–545 mS cm^{-1} at a temperature range of 30°C–120°C under 100% RH [11]. The membranes are also characterized by good thermal stability (up to 250°C), dimensional stability, and low water uptake (5.7–6.4 wt.% at 120°C). The membranes exhibit proton conductivity greater than 100 mS cm^{-1} at 120°C under 90% RH which is useful to be utilized as PEM.

Sulfonation of polymer matrix enhances the proton conductivity of the membranes, depending on the degree of sulfonation of the host polymer matrix, however, the properties of the membranes reduce beyond 200–400°C due to the progressive desulfonation of sulfonated polymers. Hence in addition to sulfonation, other functionalizations such as phosphonation and fluorination are also in recent progress for the development of fuel cell membranes. Figure 3.2 represents structures of some functionalized PEM studied for low-temperature fuel cell devices. Table 3.1 summarizes the properties of some sulfonated polymers developed in recent studies.

3.3 PHOSPHONATED PEM

Phosphonic acid groups can be the striking protogenic groups as proton conductors for intermediate temperature and low humidity fuel cell devices [24]. A high degree of self-dissociation ability promotes intrinsic amphoteric character. In addition to this, high mobility of charge carriers also eases the conduction of protons under the anhydrous condition and intermediate temperature (120–160°C). Phosphonic acid groups provide better oxidative and electrochemical stability and hence promote higher rate of reaction at the electrode surface. However, direct phosphonation of the polymer backbone is difficult in terms of control over the structure formation and minimization of undesirable side reaction which ultimately leads to insufficient material properties for the intended application. Therefore, few studies have adopted polymerization of phosphonic acid groups bearing monomers [25, 26]. In our work, we have developed a PEM based on phosphonated PEEK. The phosphonated PEEK has been synthesized by polycondensation of phosphonated bisphenol A and difluoro benzophenone [27]. Although an easy route of synthesis has been adopted for this, the degree of phosphonation is only 70%. Generally, due to the low acidity of phosphonic acid groups as compared to sulfonated groups, a high degree of phosphonation is needed to attain proton conductivity up to the desired level.

The alternative technique of phosphonation is the post-functionalization of the polymer, where an easy and simple methodology can be adopted. This method has been utilized for several aromatic hydrocarbon polymers such as polysulfone, polystyrene, polyphosphazene, polyimide, and polyphenylene oxide. Among this phosphonation of polysulfones is widely accepted due to the ease of achieving higher degree of phosphonation. A highly phosphonated polysulfone PEM has been developed by Zaidi and co-workers for DMFC applications [28]. The membrane is synthesized by post-phosphonation of polysulfones via chloromethylation followed by phosphonation via the Michaelis–Arbuzov reaction. A high degree of phosphonation of about 150% is achieved without any undesired side reactions. This high degree of functionalization leads to water uptake as high as 52% which is suitable for favorable

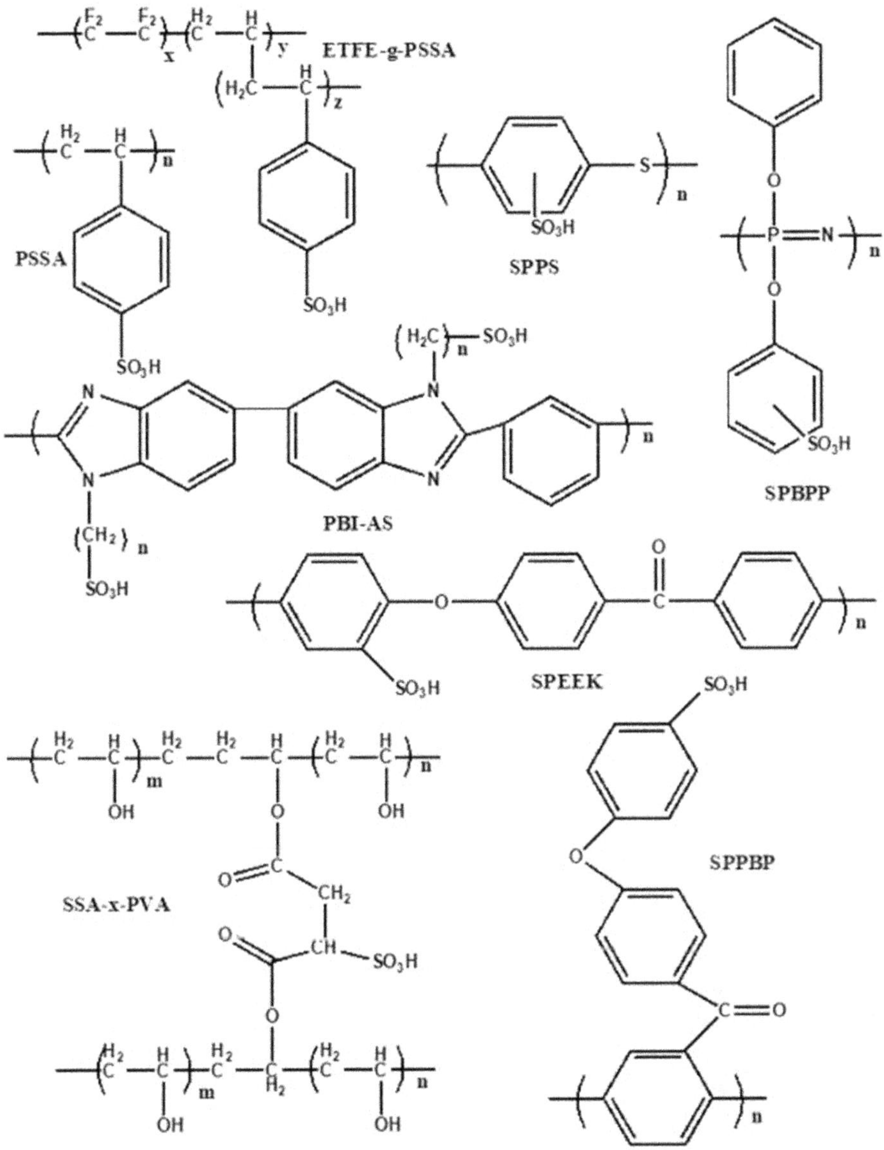

FIGURE 3.2 Chemical structures of functionalized PEM (SPEEK: Sulfonated poly(ether ether ketone), PSSA: Sulfonated polyphenylene sulfide, SSA-x-PVA: Sulfosuccinic acid crosslinked polyvinyl alcohol, SPBPP: Sulfonated poly(bisphenoxy)phosphazene, SPPS: Sulfonated Polyphenylene sulfone, PSSA: Polystyrene sulfonic acid

proton conduction process. The proton conductivity of the membranes at a 150% degree of phosphonation is 12 mS cm^{-1} at 100°C under 100% RH. Again, the thermally stable membrane has been observed to exhibit low methanol crossover (9.12 × 10^{-8} cm^2 s^{-1}) favorable in view of DMFC application. Elumalai et al. have also studied phosphonated poly(styrene-ethylene/butylene-styrene) block copolymer (PSEBS)

TABLE 3.1
Properties of sulfonated PEM

Sl no.	Membrane	Proton conductivity (S cm^{-1})	Performance	Comments	References
1	Polyethylene glycol-grafted sulfonated polyether ether ketone	10^{-5} at RT	-	Comb-like solid polymer electrolyte	[12]
2	Sulfonated poly(arylene ether sulfone) and graphene oxide composite	0.183 at 120°C and 100% RH	-	Swelling ratio reduced by 55.7% at 90°C	[13]
3	Sulfonated poly(arylene ether) multiblock copolymers	247.18 × 10^{-3} under at 80°C and 100% RH	Current density 1386 mA cm^{-2} at 0.6 V under 100% RH	Current density was 1.8 fold higher than pristine membrane	[14]
4	Cross-linked sulfonated poly(arylene ether ketone) composite membranes with silica nanoparticles	3.06 × 10^{-3} at 70°C and 30% RH	Current density 0.919 A cm^{-2} at 0.6 V under 32% RH	Improved electrochemical performance under reduced humidity	[15]
5	Crosslinked sulfonated poly(arylene ether) (cSPAE100) and triazole-functionalized poly(arylene ether sulfone)	160 × 10^{-3} at 80°C and 95% RH	Power density 0.31 W cm^{-2} at 0.82 V	Power density 1.55 times higher than cSPAE100	[16]
6	Highly branched sulfonated poly(arylene ether)s and graphene oxide	0.39 at 80°C	Power density 82.36 mW cm^{-2}	Better single cell performance than pristine membrane	[17]
7	Polybenzimidazoles bearing different acidic groups, e.g.,–SO_3H and –PO_3H_2 and amphoteric groups, e.g., imidazole and –PO_3H_2	0.015–0.075 at ambient temperatures in water 0.126 at 140°C and 30% RH	Current density 718 mA cm^{-2}	Proton conductivity was 50 times of Nafion at RH 30% Membrane exhibited excellent proton conductivity in anhydrous conditions	[18]

(Continued)

TABLE 3.1
(Continued)

Sl no.	Membrane	Proton conductivity (S cm^{-1})	Performance	Comments	References
8	Sulfonated poly(arylene ether ketone) bearing naphthalene and fluorine moieties	0.192 at 80°C	Power density 125.7 mW cm^{-2} at 80°C	Excellent single cell performance	[19]
9	Sulfonated poly(arylene ether ketone) block copolymers	131–154 × 10^{-3} at 90°C and 100% RH	Power density 324 mW cm^{-2} at 60°C under 100% RH	Morphology is related to conductivity since transport channels increased with length of hydrophilic oligomer	[20]
10	Crosslinked multiblock sulfonated poly(arylene ether ketone nitrile) with pendant 1,2,3-triazole	137–147 × 10^{-3} at 80°C and 95% RH	Power density 1.07, 0.87 and 0.48 W cm^{-2} at 80°C for 95, 70, and 30% RH	Considerable enhancement (34%–40%) of power generation	[21]
11	Highly sulfonated poly(ether ether ketone) and sulfonated PBI	0.217 at 80°C and 100% RH	Power density 171 mW cm^{-2} at RT	Power density two and half folds that of native membrane	[22]
12	Sulfonated poly(arylene ether ketone)s having photosensitive Chalcones	215 × 10^{-3} at 80°C	-	Proton conductivities similar to Nafion 117	[23]

for high temperature fuel cells [29]. The membrane reaches proton conductivity of 5.81 mS cm^{-1} at 140°C under the dry state. The membranes with a degree of phosphonation 54% show open-circuit voltage and power density of 0.77 V and 184 mW cm^{-2} at 140°C. Phosphonated polysulfones have been grafted via click chemistry in a research work [30]. Polysulfones are chosen more frequently as a base material due to the ease of functionalization coupled with good chemical, thermal, and mechanical integrity. The copolymer membranes attain conductivity as high as 83 mS cm^{-1} under immersed state at 120°C. Polyarylates bearing phosphonic acid groups are also studied for PEM applications [31]. A crosslinked poly[6-(acryloyloxy)hexylphosphonic acid membrane has been synthesized with long alkyl side chains of more than 6 methylene units. The membranes show proton conductivity comparable to standard

Nafion 117 under RH of 30%–80% at 80°C. The formation of hydrogen-bonded network structures due to the presence of aggregated hydrophilic domains is the main contributing factor for improved proton conductivity of the membrane. Phosphonated polystyrene membranes have potential as PEM material due to the high oxidative stability, low water uptake, and for their ability to show comparable proton conductivity with other phosphonated polymers used in PEM [32]. The phosphonated polymers are of recent interest as an alternative to sulfonation due to their ability of self-dissociation even at low humidity and formation of hydrogen bonding favoring easy proton transportation by structural diffusion.

3.4 FLUORINATED PEM

In addition to sulfonation and phosphonation, research on fluorinated PEM is also in the recent trend. Harit et al. in recent work have developed a fluorinated copolymer containing pyrazole side chains as a precursor for the fabrication of proton-conducting membranes for fuel cells [33]. The synthesized copolymers are thermally stable under ambient condition up to 200°C and exhibited proton conductivity as high as 4×10^{-5} mS cm^{-1} at 180°C under completely anhydrous conditions. In another work, a semi-IPN membrane based on SPEEK and crosslinked fluorinated terpolymer bearing pendant triazole groups and cyclocarbonate rings have been developed for fuel cells to be operated at low RH [34]. The semi-IPN membranes attain proton conductivity of 4.3 mS cm^{-1} at 140°C and RH of <25%. The blend membranes exhibit a glass transition temperature, tensile strength, and Young modulus of 223°C, 6%, and 4 MPa, respectively.

A combination of hydrophilic and hydrophobic polymer microstructure is advantageous for fuel cell membrane applications. Kim et al. have studied sulfonated-fluorinated copolymers synthesized by nucleophilic polycondensation [35]. The blend membrane has been prepared by a simple casting technique. With an increment in SPEEK content, the proton conduction ability of these membranes is improved. The membrane achieved a maximum conductivity value of 91 mS cm^{-1} for SFBC-50/SPEEK-20 membrane which is about 2.1 times higher than SFBC-50. The peak power density of this composition is 453.86 mW cm^{-2} which is much better than the reference membrane (Figure 3.3 (a)). The blend membrane shows good dimensional stability and current density at 70°C and 100% RH when blended with a high sulfonic acid groups containing SPEEK.

Sulfonation-fluorination combination has been introduced in the same polymer backbone to demonstrate its effect on the PEM fuel cell properties [39]. A linear polymer made of fluorinated poly(arylene ether) with and without pendant sulfonic acid groups has been synthesized from decafluorobiphenyl at room temperature. The synthesized membranes exhibit proton conductivity as high as 106 mS cm^{-1} which is higher compared to standard Nafion 212 membrane. The fluorinated membrane is also characterized by improved oxidative stability as compared to non-fluorinated membranes. In this connection, Pang et al. have developed a PEM containing hydrophilic-hydrophobic microstructures composed of fluorinated poly(arylene ether ketone) bearing pendent hexasulfophenyl groups [37]. The membranes show excellent thermal stability and mechanical strength for the desired IEC. Phase-separated morphology has been

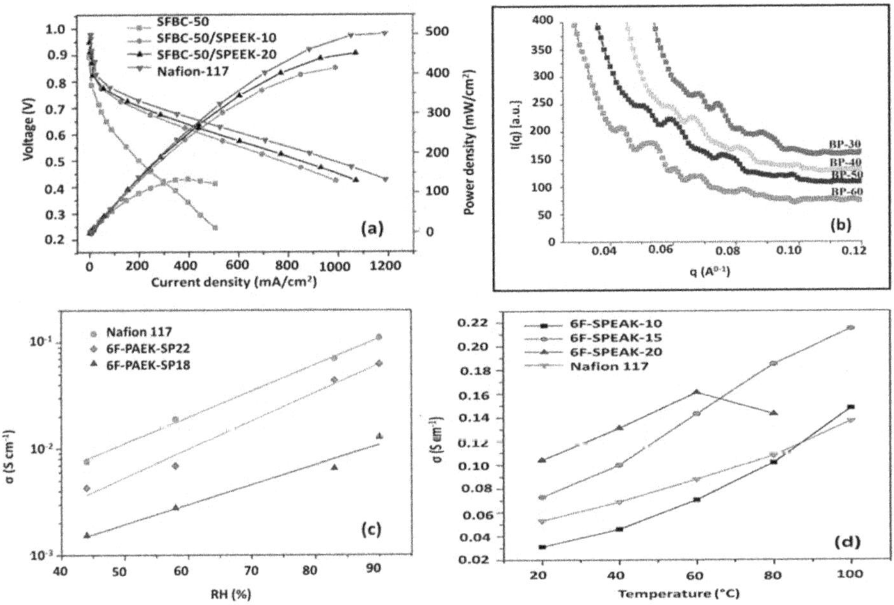

FIGURE 3.3 Performance of SFBC-50 (reference) and blend membranes at 70°C and 100% RH [35], (b) SAXS profile of sulfonated partially fluorinated polysulfone copolymers [36], (b) variation of proton conductivity with RH for fluorinated poly(arylene ether ketone) composite membrane [37] and (d) proton conductivities vs. temperature plot of hydrated 6F-SPAEK-x and Nafion 117 membranes [38] (Reprinted with permission)

utilized to obtain highly conductive and reduced swelling polymeric membranes for low-temperature fuel cells [40]. Partially fluorinated poly(arylene ether sulfone) membranes exhibit superior proton conduction ability compared to commercial Nafion 212 membrane irrespective of the level of humidification. The optimized membrane attains a current density of 488 mA cm^{-2} at 80°C under 80% RH and 0.6 V which appears to be even better than the standard Nafion 212 membrane. Xie et al. have synthesized poly(arylene ether ketone)-based membranes with octasulfonated segments by polycondensation reactions [38]. The solution-casted membranes are characterized as tough, transparent, and flexible with superb thermal and mechanical properties. The proton conductivity of these membranes attains 215 mS cm^{-1} at 100°C, which is reported to be 1.5 times better than commercial Nafion 117 under similar test conditions. The excellent combination of properties of these membranes is originated from the nanophase-separated morphology comprising hydrophobic fluorinated and hydrophilic sulfonated segments in the polymer backbone.

3.5 CARBOXYLATED PEM

Carboxylation of aromatic polymer is another approach to impart hydrophilicity in the polymer backbone. Researchers have studied carboxylated and sulfonated-carboxylated functionalization in combination and evaluated the effect of these hydrophilic groups on the proton conductivity of PEM. Fluorenyl-based poly(arylene ether) PEM has been investigated in the literature [41]. A carboxylated fluorenyl-based polymer has been post-sulfonated using conc. sulfuric acid. The polymer backbone contains three sulfonable sites for each repeating unit for the post-sulfonation process. The ion exchange capacity of these polymers can be controlled by proper optimization of the reaction time at room temperature. Kesava et al. have investigated a nanocomposite polymer membrane containing SnO_2 nanoparticles in carboxylated poly(arylene ether sulfones) matrix material [42]. A decarboxylation in PES backbone is introduced via Thiol Ene reaction mechanism. The SnO_2 nanoparticles are uniformly dispersed in this polymer matrix and exhibit considerable improvement in thermal stability along with controlled water up take and swelling that confirms the potential of this nanocomposite material to be utilized as PEM. At 3 wt.% loading of the SnO_2, the nanocomposite membrane attains proton conductivity of about 1.49 mS/cm^{-1} at 100°C. The mechanism of transport contributing to ease of proton movement inside the membrane is depicted in Figure 3.4. The mechanism proposed that the conduction might have involved both the Grotthus and Vehicle pathways. In Grotthus pathway, the proton converts carboxylated anion into carboxylated hydronium moiety. The hydronium ion then jump to the water molecules present at the close proximity which again hops and complete the proton conduction process. In Vehicular route, the water molecule present in the medium clubs with the proton and form H_3O^+ which subsequently conducts proton. The nanoparticle in the membrane helps the controlling the water up take and proton conductivity by optimization of swelling. The activation energy of the composite membrane also decreased due to favorable proton transport process.

In another interesting study, the effect of carboxylation and sulfonation in combination on the Nafion® membrane has been studied in detail to understand the shift in proton conduction mechanism with progressive increase in carboxylic functional groups. Study reveals that the proton conductivity and water permeation inside the membrane are reduced considerably above 30% carboxylation, indicating restriction in mobility of carboxylated moieties in the membrane. This observation is as expected due to low proton conduction ability of carboxylated groups as compared to highly proton-conducting sulfonated groups. Below 30% carboxylation, i.e., for 20% carboxylation water sorption is uniform and the membranes absorb much lower water compared to its sulfonated counterparts. The small angle X-ray scattering reveals

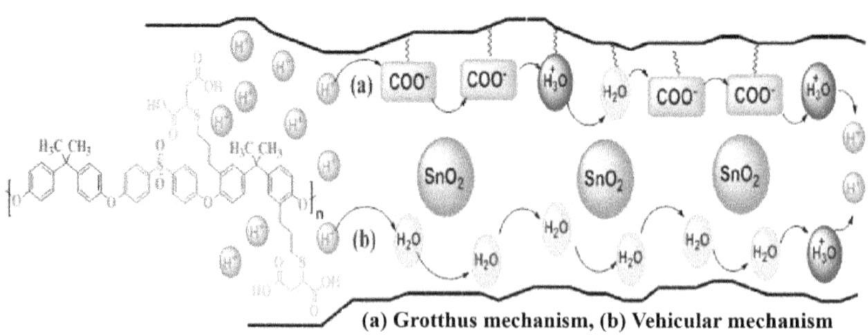

FIGURE 3.4 Proton transport process in carboxylated PEM [42] (Reprinted with permission)

that the composites with less carboxyl groups form ionic domains of size comparable to the Nafion®. Proton transport mechanism inside the membrane is altered with the change in carboxylic acid groups. The carboxylic acid-rich phase in the composite exhibits short and strong hydrogen bonding due to less acidity of carboxylic groups as compared to sulfonic acid groups. Hence, it is expected that with an increase in the –COOH groups, the mobility of proton is reduced and the mechanism is now shifted from vehicular to Grotthuss way. According to the study, with an increment in carboxylate functional groups the water uptake is lowered however, water sorption study exhibits that water is much more tightly bound to composite membranes as compared to Nafion® in carboxylic-containing membranes.

Again carboxylated functionalities are introduced to impart crosslinking or grafting in benzimidazole type of polymers to control the swelling behavior by maintaining desired proton conductivity [43]. Liu et al. have prepared polyarylene ether ketones containing sulfonated and carboxylated groups and PVA by crosslinking approach [41]. The PEM so synthesized obtained proton conductivity of 0.15 S/cm^{-1} with desired dimensional stability. The stability of the membranes has been further enhanced due to the presence of –COOH groups known for making strong hydrogen bonding between the two polymers.

3.6 SIGNIFICANCE OF FUNCTIONALIZED PEM

As discussed, functionalization remains an effective tool to improve the proton conduction ability of the base polymers by judicially introducing hydrophilic domains inside the polymer backbone. The research is focused on the development of sulfonated hydrocarbon based polymers with aromatic backbones due to excellent combination of chemical, thermal, and mechanical integrity. These polymers are easy to synthesize specifically, the sulfonated one and are robust and less costly. The aromatic nature of the polymer backbone helps in formation of intramolecular π-π interaction due to probable stacking of the layers. This in turn provides enhanced mechanical and chemical stability with uniform film formation ability. However, the designing of the PEM is to be done in such a way that the proton conductivity and mechanical integrity can be balanced as per requirement.

3.7 LIMITATION OF FUNCTIONALIZED PEM

Although, sulfonation remains an effective strategy however, it is not always straightforward since increase in degree of functionalization raises the water uptake and swelling and therefore maintaining required mechanical stability is challenge. A very high degree of functionalization generates drastic drop in dimensional stability and the high swelling tendency of the membranes in water is also a matter of concern for long-term durability of the PEM. The membrane materials suffer from huge swelling and gel type structure is formed in wet condition whereas as the membranes get dried it is converted in to a brittle structure. Hence, to balance the drop of dimensional stability, one usual practice is to impart crosslinking inside the polymer chains. However, % crosslinking needs to be controlled to handle the brittleness of the membrane and drop in proton conduction property. Crosslinking reduces water uptake and extent of swelling however, reports revealed that a high level of crosslinking leads to hardness and brittleness in the membrane with decrease in proton conductivity. Hence, a judicial balance between the two is essential to achieve practically usable proton conductive and mechanically stable membrane.

3.8 CHAPTER SUMMARY

The microstructural investigation of commercial PEM depicts that presence of hydrophilic and hydrophobic structures are required for the formation of efficient proton conduction path. The hydrophilic groups in alternative PEM are often introduced by functionalization approach. Post-sulfonation remains the easiest and most explored approach for the synthesis of sulfonated PEM. In addition to this, researcher explored the effect of fluorination, phosphonation, carboxylation of polymer matrix in the proton conductivity of membrane as well. The pre-functionalization of PEM has also been done to obtain uniform distribution of functional groups inside the matrix. However, extensive functionalization of hydrophilic moiety leads to drastic drop in dimensional stability hence optimization is required to achieve highly conductive yet durable PEM. In this chapter, we have discussed recent development in different types of functionalized PEM utilized for low temperature fuel cell application.

REFERENCES

[1] Banerjee S, Kar K, Das M. Electrolyte membranes for fuel cells: Synthesis, characterization and degradation analysis. *Recent Pat Mater Sci* 2014;7:173–203. https://doi.org/10.2174/1874464807666141001231351.

[2] Banerjee S, Kar KK. Impact of degree of sulfonation on microstructure, thermal, thermomechanical and physicochemical properties of sulfonated poly ether ether ketone. *Polymer (Guildf)* 2017;109:176–86. https://doi.org/10.1016/j.polymer.2016.12.030.

[3] Banerjee S, Kar KK. Aluminum-substituted phosphotungstic acid/sulfonated poly ether ether ketone nanocomposite membrane with reduced leaching and improved proton conductivity. *High Perform Polym* 2016;28:1043–58. https://doi.org/10.1177/0954008315614984.

[4] Banerjee S, Kar KK. Superior water retention, ionic conductivity and thermal stability of sulfonated poly ether ether ketone/polypyrrole/aluminum phosphate nanocomposite based polymer electrolyte membrane. *J Environ Chem Eng* 2016;4:299–310. https://doi.org/10.1016/j.jece.2015.11.033.

[5] Banerjee S, Kar KK. Synergistic effect of aluminium phosphate and tungstophosphoric acid on the physicochemical properties of sulfonated poly ether ether ketone nanocomposite membrane. *J Appl Polym Sci* 2016;133:42952–62. https://doi.org/10.1002/app.42952.

[6] Bae I, Oh K-H, Yun S-H, Kim H. Asymmetric silica composite polymer electrolyte membrane for water management of fuel cells. *J Memb Sci* 2017;542:52–9. https://doi.org/10.1016/j.memsci.2017.07.058.

[7] Higashihara T, Matsumoto K, Ueda M. Sulfonated aromatic hydrocarbon polymers as proton exchange membranes for fuel cells. *Polymer (Guildf)* 2009;50:5341–57. https://doi.org/10.1016/j.polymer.2009.09.001.

[8] Song M, Lu X, Li Z, Liu G, Yin X, Wang Y. Compatible ionic crosslinking composite membranes based on SPEEK and PBI for high temperature proton exchange membranes. *Int J Hydrogen Energy* 2016;41:12069–81. https://doi.org/10.1016/j.ijhydene.2016.05.227.

[9] Anis A, Al-Zahrani SM. Sulfonated PVA / PBI based crosslinked composites towards anhydrous proton conductive polymer electrolyte membranes for fuel cells. *Int J Electrochem Sci* 2012;7:9174–85. https://doi.org/10.1016/S1452-3981(23)16189-X.

[10] Ahmed F, Sutradhar SC, Ryu T, Jang H, Choi K, Yang H, et al. Comparative study of sulfonated branched and linear poly(phenylene)s polymer electrolyte membranes for fuel cells. *Int J Hydrogen Energy* 2018;43:5374–85. https://doi.org/10.1016/j.ijhydene.2017.08.175.

[11] Feng S, Sasaki K, Nishihara M. Effect of sulfonation level on sulfonated aromatic poly(ether sulfone) membranes as polymer electrolyte for high-temperature polymer electrolyte membrane fuel cells. *Macromol Chem Phys* 2016;217:2692–9. https://doi.org/10.1002/macp.201600397.

[12] Guo M, Zhang M, He D, Hu J, Wang X, Gong C, et al. Comb-like solid polymer electrolyte based on polyethylene glycol-grafted sulfonated polyether ether ketone. *Electrochim Acta* 2017;255:396–404. https://doi.org/10.1016/j.electacta.2017.10.033.

[13] Chen R, Xu F, Fu K, Zhou J, Shi Q, Xue C, et al. Enhanced proton conductivity and dimensional stability of proton exchange membrane based on sulfonated poly(arylene ether sulfone) and graphene oxide. *Mater Res Bull* 2018;103:142–9. https://doi.org/10.1016/j.materresbull.2018.02.036.

[14] Yoo T, Aziz Md A, Oh K, Shanmugam S. Modified sulfonated Poly(arylene ether) multiblock copolymers containing highly sulfonated blocks for polymer electrolyte membrane fuel cells. *J Memb Sci* 2017;542:102–9. https://doi.org/10.1016/j.memsci.2017.08.008.

[15] Bae I, Oh K-H, Yun M, Kang MK, Song HH, Kim H. Nanostructured composite membrane with cross-linked sulfonated poly(arylene ether ketone)/silica for high-performance polymer electrolyte membrane fuel cells under low relative humidity. *J Memb Sci* 2018;549:567–74. https://doi.org/10.1016/j.memsci.2017.12.060.

[16] Ahn M-K, Lee B, Jang J, Min C-M, Lee S-B, Pak C, et al. Facile preparation of blend proton exchange membranes with highly sulfonated poly(arylene ether) and poly(arylene ether sulfone) bearing dense triazoles. *J Memb Sci* 2018;560:58–66. https://doi.org/10.1016/j.memsci.2018.05.011.

[17] Liu D, Peng J, Li Z, Liu B, Wang L. Improvement in the mechanical properties, proton conductivity, and methanol resistance of highly branched sulfonated poly(arylene ether)/graphene oxide grafted with flexible alkylsulfonated side chains nanocomposite membranes. *J Power Sources* 2018;378:451–9. https://doi.org/10.1016/j.jpowsour.2017.12.057.

[18] Firouz Tadavani K, Abdolmaleki A, Molavian MR, Borandeh S, Sorvand E, Zhiani M. Synergistic behavior of phosphonated and sulfonated groups on proton conductivity and their performance for high-temperature proton exchange membrane fuel cells (PEMFCs). *Energy & Fuels* 2017;31:11460–70. https://doi.org/10.1021/acs.energyfuels.7b01065.

[19] Ru C, Li Z, Zhao C, Duan Y, Zhuang Z, Bu F, et al. Enhanced proton conductivity of sulfonated hybrid poly(arylene ether ketone) membranes by incorporating an amino–sulfo bifunctionalized metal–organic framework for direct methanol fuel cells. *ACS Appl Mater Interfaces* 2018;10:7963–73. https://doi.org/10.1021/acsami.7b17299.

[20] Lee KH, Chu JY, Kim AR, Yoo DJ. Enhanced performance of a sulfonated poly(arylene ether ketone) block copolymer bearing pendant sulfonic acid groups for polymer electrolyte membrane fuel cells operating at 80% relative humidity. *ACS Appl Mater Interfaces* 2018;10:20835–44. https://doi.org/10.1021/acsami.8b03790.

[21] Hu H, Dong T, Sui Y, Li N, Ueda M, Wang L, et al. A thermally crosslinked multiblock sulfonated poly(arylene ether ketone nitrile) copolymer with a 1,2,3-triazole pendant for proton conducting membranes. *J Mater Chem A Mater* 2018;6:3560–70. https://doi.org/10.1039/C7TA10290A.

[22] Gao S, Chen X, Xu H, Luo T, Ouadah A, Fang Z, et al. Sulfonated graphene oxide-doped proton conductive membranes based on polymer blends of highly sulfonated poly(ether ether ketone) and sulfonated polybenzimidazole. *J Appl Polym Sci* 2018;135:46547–57. https://doi.org/10.1002/app.46547.
[23] Yan S, Zhong Z, Li X, Wen P, Zhang H, Li L, et al. Photocrosslinking of sulfonated poly(arylene ether ketone)s in a hydrated state to obtain proton exchange membranes with high performance. *Macromol Chem Phys* 2018;219:1800056–65. https://doi.org/10.1002/macp.201800056.
[24] Schuster M, Rager T, Noda A, Kreuer KD, Maier J. About the choice of the protogenic group in pem separator materials for intermediate temperature, low humidity operation: A critical comparison of sulfonic acid, phosphonic acid and imidazole functionalized model compounds. *Fuel Cells* 2005;5:355–65. https://doi.org/10.1002/fuce.200400059.
[25] Fukuzaki N, Nakabayashi K, Nakazawa S, Murata S, Higashihara T, Ueda M. Highly phosphonated poly(*N*- phenylacrylamide) for proton exchange membranes. *J Polym Sci A Polym Chem* 2011;49:93–100. https://doi.org/10.1002/pola.24422.
[26] Yamabe M, Akiyama K, Akatsuka Y, Kato M. Novel phosphonated perfluorocarbon polymers. *Eur Polym J* 2000;36:1035–41. https://doi.org/10.1016/S0014-3057(99)00158-5.
[27] Banerjee S, Kar KK, Ghorai MK, Das S. Synthesis of polyether ether ketone membrane with pendent phosphonic acid group and determination of proton conductivity and thermal stability. *High Perform Polym* 2015;27:402–11. https://doi.org/10.1177/0954008314553643.
[28] Abu-Thabit NY, Ali SA, Javaid Zaidi SM. New highly phosphonated polysulfone membranes for PEM fuel cells. *J Memb Sci* 2010;360:26–33. https://doi.org/10.1016/j.memsci.2010.04.041.
[29] Elumalai V, Annapooranan R, Ganapathikrishnan M, Sangeetha D. A synthesis study of phosphonated PSEBS for high temperature proton exchange membrane fuel cells. *J Appl Polym Sci* 2018;135:45954–63. https://doi.org/10.1002/app.45954.
[30] Dimitrov I, Takamuku S, Jankova K, Jannasch P, Hvilsted S. Polysulfone functionalized with phosphonated poly(pentafluorostyrene) grafts for potential fuel cell applications. *Macromol Rapid Commun* 2012;33:1368–74. https://doi.org/10.1002/marc.201200216.
[31] Higashihara T, Fukuzaki N, Tamura Y, Rho Y, Nakabayashi K, Nakazawa S, et al. Polymer electrolyte membrane based on polyacrylate with phosphonic acidvia long alkyl side chains. *J Mater Chem A* 2013;1:1457–64. https://doi.org/10.1039/C2TA00537A.
[32] Tamura Y, Sheng L, Nakazawa S, Higashihara T, Ueda M. Polymer electrolyte membranes based on polystyrenes with phosphonic acid via long alkyl side chains. *J Polym Sci A Polym Chem* 2012;50:4334–40. https://doi.org/10.1002/pola.26246.
[33] Harit T, Malek F, Ameduri B. Fluorinated polymers based on pyrazole groups for fuel cell membranes. *Eur Polym J* 2016;79:72–81. https://doi.org/10.1016/j.eurpolymj.2016.04.017.
[34] Campagne B, David G, Améduri B, Jones DJ, Rozière J, Roche I. New semi-IPN PEMFC membranes composed of crosslinked fluorinated copolymer bearing triazole groups and sPEEK for operation at low relative humidity. *Int J Hydrogen Energy* 2015;40:16797–813. https://doi.org/10.1016/j.ijhydene.2015.08.052.
[35] Kim AR, Vinothkannan M, Yoo DJ. Sulfonated-fluorinated copolymer blending membranes containing SPEEK for use as the electrolyte in polymer electrolyte fuel cells (PEFC). *Int J Hydrogen Energy* 2017;42:4349–65. https://doi.org/10.1016/j.ijhydene.2016.11.161.
[36] Mohammadi M, Mehdipour-Ataei S. Durable sulfonated partially fluorinated polysulfones as membrane for PEM fuel cell. *Renew Energy* 2020;158:421–30. https://doi.org/10.1016/j.renene.2020.05.124.

[37] Pang J, Jin X, Wang Y, Feng S, Shen K, Wang G. Fluorinated poly(arylene ether ketone) containing pendent hexasulfophenyl for proton exchange membrane. *J Memb Sci* 2015;492:67–76. https://doi.org/10.1016/j.memsci.2015.05.027.

[38] Xie Y, Liu D, Li D, Han X, Li S, Chen Z, et al. Highly proton conducting proton-exchange membranes based on fluorinated poly(arylene ether ketone)s with octasulfonated segments. *J Polym Sci A Polym Chem* 2018;56:25–37. https://doi.org/10.1002/pola.28857.

[39] Xu J, Lin Q, Yu Y, Chen D, Ye Z. Facile synthesis of fluorinated poly(arylene ether)s with pendant sulfonic acid groups for proton exchange membranes. *Int J Hydrogen Energy* 2017;42:27100–10. https://doi.org/10.1016/j.ijhydene.2017.09.032.

[40] Lee S, Yuk J, Nugraha AF, Shul Y, Park S, Shin D, et al. Partially fluorinated multiblock poly(arylene ether sulfone) membranes for fuel cell applications. *Macromol Mater Eng* 2018;303:1700650–4. https://doi.org/10.1002/mame.201700650.

[41] Liu B, Hu W, Robertson GP, Guiver MD. Poly(aryl ether ketone)s with carboxylic acid groups: Aynthesis, sulfonation and crosslinking. *J Mater Chem* 2008;18:4675–82. https://doi.org/10.1039/b806690f.

[42] Kesava M, Dinakaran K. SnO_2 nanoparticles dispersed carboxylated Poly(arylene ether sulfones) nanocomposites for proton exchange membrane fuel cell (PEMFC) applications. *Int J Hydrogen Energy* 2021;46:1121–32. https://doi.org/10.1016/j.ijhydene.2020.09.178.

[43] Lin H, Zhao C, Cui Z, Ma W, Fu T, Na H, et al. Novel sulfonated poly(arylene ether ketone) copolymers bearing carboxylic or benzimidazole pendant groups for proton exchange membranes. *J Power Sources* 2009;193:507–14. https://doi.org/10.1016/j.jpowsour.2009.04.036.

4 Acid-base complex-based PEM

4.1 ANHYDROUS PEM

Polymer electrolyte membrane fuel cell (PEMFC) is operational at considerably lower temperature compared to the other fuel cells and delivers power output suitable for stationary, automobile, and portable devices. The success of PEMFC is much dependent on the hydrogen infrastructure, i.e., generation, storage, transportation, and handling, for the large-scale production. The reformate hydrogen contains CO and CO_2 that are poisonous for the Pt catalyst. At temperature of around 80°C, the tolerance of CO is poor and needs removal of CO from the reformate hydrogen prior to utilization in fuel cell devices. The CO poisoning is a temperature governing process and a considerable tolerance can be achieved at a temperature range of about 150–200°C. This demands the shift of PEM research focus toward high-temperature anhydrous membranes based on azole, triazole, ionic liquid, etc. The high temperature fuel cells are equipped with simple reformers. The fuel cell reactions are exothermic in nature and the heat generated in the process can be essentially utilized for the purpose of preheating. However, the utilization of waste heat is possible as long as the temperature of reformer and cell stack are comparable. The integrated power system can be developed which can operate at higher temperature of around 240–300°C with efficiently designed compact system. In the subsequent section, we will discuss the recent development in the field of anhydrous PEM which are suitable for fabrication of high temperature fuel cells.

4.2 PROTON CONDUCTION IN ANHYDROUS PEM

In general, fuel cell membranes exhibit water-dependent proton conduction behavior. The water-assisted conduction mechanism demands a sufficient amount of hydration level to achieve desirable proton conductivity. This renders the use of these membranes for high-temperature applications preferably over 100°C. Sulfonated polymeric membranes hold few water molecules per $-SO_3H$ groups even over the boiling point of water, indicating no true anhydrous state of those membranes [1]. However, this small amount of water is not sufficient to make a continuous proton conduction pathway and hence demands the development of new membranes. Nitrogen-containing bases such as imidazoles, imidazole derivatives, triazoles, and their derivatives have been used as a water substitute. These groups of compounds are commonly known as proton transport facilitators that exist in tautomeric form. Azoles are typically amphoteric in nature and incorporation of them into the polymers with Bronsted acid functionality is known to produce high proton conductivity.

DOI: 10.1201/9781003733744-4

The chemical properties of imidazole and pyrazole groups of materials empower them to act as a proton donor and acceptor during the conduction process. In this respect, these compounds are amphoteric in nature however in the presence of other compounds they are more basic as compared to the water. Imidazoles act as a proton solvent and form an intermolecular hydrogen bonding and self-dissociate in the presence of a Bronsted acid, such as sulfonated polymers. The basic nitrogen atom plays the role of a proton acceptor in the presence of sulfonic acid groups by forming protonic charge carriers, $(C_3H_3NH_2)^+$. Nitrogen-containing bases are commonly incorporated by doping with sulfonated polymers via the physical absorption of azoles in the host polymer matrix. However, the main disadvantage of this method remains that the azoles are dragged out of the membrane at high humidity due to the extensive solubility in water. Flexible spacers are often used to bind the azoles into the polymeric backbones. Doping with a small amount of acid conveys mobile excess protons into the system. It is to be understood that, methodically this process of doping is different than the swelling of the polymers such as PBI in the presence of the phosphoric acid (PA). In the case of later, the proton conduction is dependent on the bulk properties of the PA, whereas in the case of the earlier, acid plays the role of a proton source instead of a proton solvent [2]. In Figure 4.1 we have represented chemical structures of acidic and basic polymers.

4.3 PBI-BASED ACID-BASE COMPLEXES

PBI membranes are explored in different research fields due to high ionic conductivity in anhydrous condition, good thermal stability, excellent physical properties and robust mechanical integrity, and low fuel crossover. All of these contribute to the application of PBI in high-temperature fuel cells, anionic fuel cells, alkaline fuel cells, batteries, desalination, water treatment, etc. [4–6]. The functionalized form of PBI has also been extensively explored in growing fields like sensors, actuators, flow batteries, hydrogen purification, and many more [7, 8]. As discussed earlier, the major shortcoming of PEMs is that they have to be properly hydrated for efficient functioning. This limits application of these PEMs for high-temperature fuel cell devices and hence requires proper water management. Two different approaches are taken to resolve these issues. First, to develop acid-base complexes composed of a basic polymer such as PEO, PVP, PVA, and PEI, and an acid. Polymers having functional groups such as imine, amide, ether, and imide, produce ion pairs with strong acids by the acid-base reaction. PBI doped phosphoric acid is the most accepted example of the class. PBI acts as the basic polymer characterized by excellent chemical and mechanical stability at an operating temperature of 120–200°C. H_3PO_4 as the acidic part of the complex supports the process of proton conduction in water-free condition. Upon doping PBI with strong acids like phosphoric acid, acid-base interaction between imidazolium groups of PBI and acid molecules takes place via hydrogen bonding which, in turn, reduces the cohesive forces between the polymer chains. This interaction is the main responsible factor for proton conduction in the PBI matrix. Based on the structural arrangement in PBI, at most four molecules of the phosphoric acid may be protonated while the rest of the acidic parts remain un-dissociated. The excess acids remain therein form hydrogen bonds

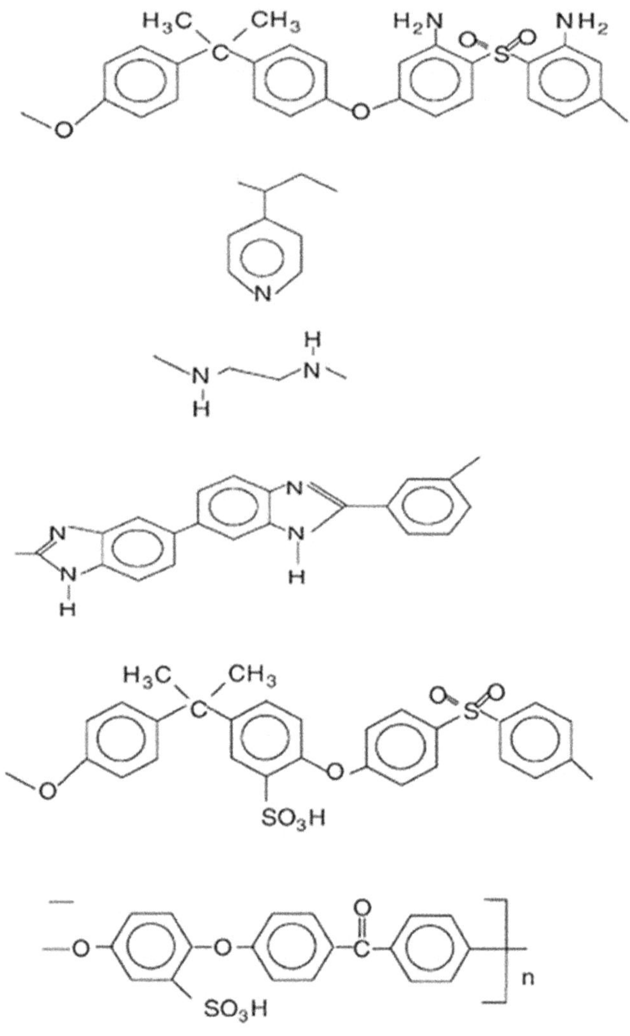

FIGURE 4.1 Chemical structures of common acidic and basic polymer forming acid-base complex PEM [3] (Reprinted with permission)

and conduct protons. The membranes can successfully operate without external humidification by diffusion of water generated at the cathode side. The second approach uses a polymer doped with ionic liquid. The incorporation of ionic liquid enables the use of membranes for medium temperature fuel cell devices due to their high boiling points (>100°C). Ionic liquids serve the role of proton conductor due to the presence of protons in their structure which can form hydrogen bonding. The addition of them improves the proton conductivity and membrane exhibits proton conductivity as high as 10^{-1} to 10^{-2} S cm^{-1} at 160°C [9, 10]. The variation in proton conductivity for ionic liquid doped composite membranes with temperature is represented in Figure 4.2.

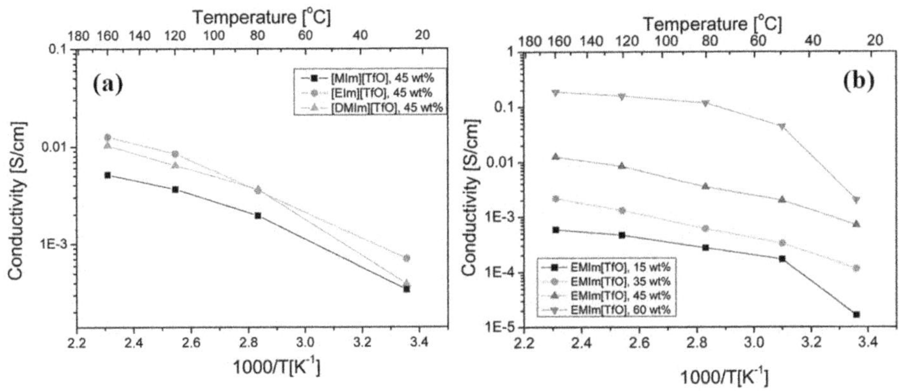

FIGURE 4.2 (a) Proton conductivity for PIL based polymer composite and (b) [EIm][TfO]/polymer composite with variation in temperature in the form of Arrhenius plot [10] (Reprinted with permission)

The presence of PBI allows higher working temperature for fuel cell devices. Seel et al. prepared phosphoric acid doped PBI membranes by synthesizing PBI in polyphosphoric acid and then casted in membrane form [11]. The homogeneous membranes so prepared are characterized by good proton conduction properties and mechanical stability. However, the drawbacks of phosphoric acid doped PBI open further areas of research. The major problems remain in (a) the water formed at the cathode side leads to leaching of the phosphoric acid with time and (b) reduction in mechanical stability of the complex membranes with progressive addition of phosphoric acid. When the membrane comes in contact with the water vapor, phosphoric acid leaches out of the membrane which in turn lowers the proton conductivity and accelerates corrosion. Different attempts have been taken to overcome these problems, crosslinking is one of the most commonly used strategy. Theoretically, various crosslinking approaches can be adopted to reduce these problems either by ionic crosslinking or covalent crosslinking or formation of organic-inorganic hybrid membranes. Devrim et al. have developed crosslinked PEM based on PBI for high-temperature fuel cells [12]. The crosslinking reaction is performed with different crosslinkers such as ethylene glycol diglycidyl ether (EGDE), a-a′-dibromo-p-xylene (DBpX), bisphenol A diglycidyl ether (BADGE), and terephthaldehyde (TPA). The retention of acid is improved by crosslinking and maximum acid retention properties have been observed for PBI/BADGE membrane. PBI/BADGE membrane exhibits a current density of 0.121 A cm^{-2} at an operating voltage of 0.6 V. The membrane also has attained a power density of 0.123 W cm^{-2}. A heavily phosphoric acid doped (30–35 mol PA/PBI) high molecular weight PBI membrane has been developed by Mader and Benicewicz for fuel cell devices [13]. The proton conductivity of these membranes remains between 0.1 and 0.25 S cm^{-1} at 180°C depending on the level of acid doping. These membranes are promising for high-temperature fuel cells and show a degradation rate of 30 μV h^{-1} in an initial performance stability test for 3000 hours.

To overcome the associated challenges of acid-base complex systems, an alternative class of proton-conducting membrane has been introduced by Kreuer [14]. The new system contains an acidic polymer and nitrogen-containing heterocycle serving the role of a proton solvent. Hence, performance of this system is independent of the humidity issue. The heterocycles function in a similar way like water and can form hydrogen-bonded structure by undergoing autoprotolysis to a high extent leading to excellent proton conduction properties [15]. In addition, the second major component of the system, i.e., the host polymer matrix maintains desired mechanical strength enabling film formation with great stability. Again, the host polymers serve as the source of protons, supply protons to the heterocycles, and also act as the protogenic solvents. The heterocycles, in general, have high boiling points which make them suitable for high-temperature fuel cell applications. Unlike water, they can be added in the polymer matrix to lower the volatility further during the operation. In these systems, nitrogen groups of the heterocycles play the role of proton acceptor against the strong acidic groups such as sulfonic or phosphonic acid groups of the polymer backbone. Moreover, the heterocycles undergo self-dissociation at a much higher extent as compared to the pure water leading to high proton mobility and conductivity.

4.4 AZOLE-BASED HETEROCYCLES

4.4.1 Imidazole (Im)

One of the most explored heterocycles for fuel cell application is imidazole (Im). In respect to physical property, Im is a solid material having a very high boiling point of 257°C. It exhibits a proton conductivity value of 10^{-3} S cm^{-1} at its melting point. Im has pKa value of 14.9 and the proton conduction process in Im-based systems is governed by the Grotthuss mechanism, i.e., by conduction of proton between the protonated and unprotonated molecules [16]. A proposed conduction mechanism in Im is shown in Figure 4.3. The conduction mechanism involves proton transfer between the adjacent Im molecules via the reorientation of the Im molecules. This reorientation process is considered as the rate-limiting factor for long-range proton conduction across the Im-based membranes. Quantum mechanical calculation on a nitrogen-based system reveals that there is a possibility of hydrogen bond formation between nitrogen molecules which lowers the proton transfer barrier [17].

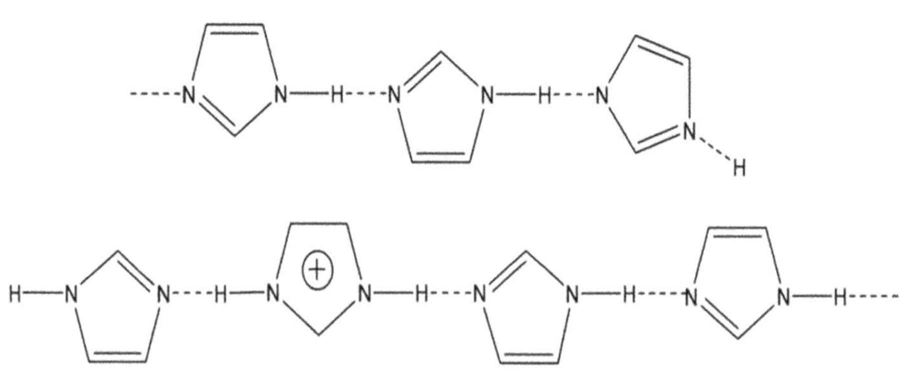

FIGURE 4.3 Proton conduction mechanism in Im via the process of reorientation

4.4.2 Benzimidazole (BnIm)

Benzimidazole (BnIm) is another heterocyclic compound made by the fusion of benzene and imidazole. The proton conduction process in BnIm is also governed by the Grotthuss mechanism similar to Im. During the process of conduction, the protonated and unprotonated nitrogen molecules act as donor and acceptor molecules in the transfer reaction. BnIm has a pKa value of 12.9, and the value is slightly lower than the Im, however, BnIm can be successfully used as dopant in acidic polymeric materials to ease the conduction process via Grotthuss pathways [16]. In addition to this, BnIm also has added advantage of electrochemical stability which makes them suitable for fuel cell application.

4.4.3 Triazole (Tri)

Triazole is another important class of heterocyclic compound which can be effectively incorporated into the acidic host polymer matrices to improve proton conductivity in dry condition. 1H-1,2,3 triazole is liquid at room temperature having melting and boiling point of 23 and 203°C [18]. The chemical structure of Tri is similar to that of other heterocycles, e.g., Im and Py, with a distinct difference of the presence of three nitrogen atoms in a five-membered ring structure of Tri. Hence, the proton conduction process in Tri is similar to that of other heterocycles. However, in a research article Zhou et al. have suggested that the proton conduction in Tri based matrices is different as compared to Im due to the presence of numerous proton-carrying isomers [17].

4.4.4 Polyacid/azole blends

The above-mentioned heterocycles are blended with acidic polymers to explore their proton conduction property. The acidic functional groups present in the polymers are one of the responsible factors for easy proton conduction process. Several polymers with different functional moieties such as carboxylic, sulfonic and phosphonic acid groups are studied in the literature [19–21].

4.4.4.1 Polyacids containing carboxylic groups

Carboxyl group-containing polyacids inherently suffer from the problem of lower sensitivity toward hydrolysis and high pKa value, which eventually leads to the low proton conductivity. To address these issues, in some recent research works, attempts have been made to prepare PEMs for fuel cells by doping with heterocycle compounds. Karadedeli et al. have developed PEMs based on adipic acid and BnIm at different molar concentrations [22]. Although the proton conductivity value of adipic acid is quite low ($\sim 10^{-11}$ S cm^{-1}) in its pure form however the conductivity attains an impressive value of 4×10^{-3} S cm^{-1} at 130°C when added with BnIm. The conductivity of the electrolyte system is increased with an increase in BnIm content. The study reveals that the development of protonic defect serves as a solvent for protons. In another study, Yamada and Honma have studied a blend consisting of alginic acid and Im as heterocycle [23]. The membranes are flexible, free-standing, transparent, and thermally stable up to 140°C for Im to carboxylic group ratio from 0 to 5. The study reports proton conductivity of 2×10^{-3} S cm^{-1} at 130°C and dry state for the molar ratio of Im to –COOH ≥ 2, however, the proton conductivity is reduced and attains a value of (< 10^{-8} S cm^{-1}) for molar ratio < 2.

4.4.4.2 Polyacids containing phosphonic acid groups

Polyacids with phosphonic acid functionalization are also among the material of choice for the fuel cell applications. Lee et al. have studied the structure and proton mobility of poly(vinyl phosphonic acid) (PVPA) with the help of solid-state NMR [24]. The results reveal that the proton migration in PVPA is assisted by the hydrogen-bonded network formed by the acidic protons. However, the condensation of acidic protons inhibits the proton transport and hence a significant reduction in proton conductivity is observed. Study reveals that annealing and drying of the material significantly affect the water content of the polymer and hence the proton conductivity as water content plays a dominant role in the proton conduction mechanism. Hence, condensation plays a crucial role in proton conductivity as it reduces the availability of phosphonic acid groups for the conduction process. PVPA/BnIm doped polymer electrolyte membranes with various doping levels ranging from 0.5 to 2 have been developed by Sevil et al. [25]. Based on IR studies, it is observed that the heterocycle units are protonated from the free nitrogen sides. The membranes developed are characterized by good proton conductivity (10^{-3} S cm^{-1} at 150°C for x = 2) and thermally stability up to 200°C. Glass filter supported BnIm and monododecyl phosphate (MDP) blend have been studied by Yamada and Honma [26]. The thermally stable membranes attain proton conductivity as high as 1×10^{-3} S cm^{-1} at 150°C for BnIm/MDP weight ratio of 200 wt.%. However, the proton conductivity of these membranes declines at and above 120°C due to the evaporation of MDP from the membrane. Gunday et al. have reported a blend membrane based on 1H-1,2,4-triazole and PVPA to demonstrate the interaction between the two material followed by its effect on the device performance [20]. The electrolytes are thermally stable up to 150°C and exhibit a maximum proton conductivity of 3×10^{-3} S cm^{-1} at 120°C. The proton conductivity of the electrolyte has been observed to be improved with an increase in triazole content. This improvement in proton conductivity has been attributed to the increment in unprotonated nitrogen site groups similar to Im. The proton conduction process in Tri/PVPA blend is governed by the mobility of protonic defects by structural diffusion, i.e., by Grotthuss pathways. In addition to this, authors also suggest that the formation of dynamic hydrogen-bonded chain structure via ionic aggregation may result in long-range proton conduction channels which, in turn, favor high proton conductivity.

4.4.4.3 Polyacids containing sulfonic acid groups

Polystyrene sulfonic acid (PSSA) is the common polyacid containing sulfonic acid functionality, obtained by sulfonating commercial polystyrene [27]. PSSA in its hydrated form exhibits good proton conductivity which makes it suitable material for PEM fuel cells however, the major limitation lies in their instability. PSSA suffers from oxidative decomposition due to the attack of the oxygen radicals obtained at the cathode side of the membrane electrode assembly [28]. On this basis, PSSA has been doped with Tri and Im containing triethylene glycol to fabricate a low cost thermally stable highly proton conductive membrane [27]. The composite membranes are fabricated by doping in PSSA at various molar ratios. The proton conductivity attains a maximum of 5×10^{-4} S cm^{-1} at room temperature. The proton conduction in this system has been executed by the Grotthuss mechanism utilizing structural diffusion via intermolecular proton transport between the terminating Im units. Again, the proton conductivity of PSSA/Tri composite has been five times higher as compared to PSSA/BnIm doped membranes at room temperature. Sulfonated polysulfone (sPSU) has been another attractive PEM material due to the mechanical strength and low gas permeation. These membranes can be effectively used as a high-temperature PEM material since the introduction of sulfonic acid groups improves thermal and mechanical stability and glass transition temperature of the host polymer matrix. Zhou et al. have studied sPSU/triazole solid polymeric membranes at different triazole concentration [18]. The membrane exhibits conductivity as high as 0.01 S cm^{-1} at 110°C under anhydrous condition. The protons conduction mechanism in these membranes takes place either through a long-range motion of the triazole moieties or by intermolecular proton transport among the adjacent Tri groups.

4.5 IONIC LIQUID-DOPED PEM

Ionic liquids are highly conductive organic salts with a melting point around or below the room temperature. They are characterized by an excellent combination of properties such as high ionic conductivity under anhydrous condition, low volatility, high thermal and electrochemical stability, and non-flammability. Ionic liquids exhibit conductivity in the order of 10^{-4} to 10^{-2} S cm^{-1} at room temperature. Based on chemical structure, the ionic liquid can be classified into two classes, polar protic, and aprotic ones. The aprotic ionic liquids are most preferably used for the lithium-ion batteries due to the high-ionic mobility and concentrations. The protic ionic liquids are selectively used for fuel cell devices owing to the presence of mobile protons situated on the cation. The chemical reactivity of these active protons makes them suitable for fuel cell devices.

4.5.1 IONIC LIQUID AS ELECTROLYTE

Ionic liquids are prepared by combining Bronsted acids and bases. The Bronsted acid-base combination transfers protons from acid to the base via hydrogen-bonded network formation due to the presence of proton donor-acceptor sites. In addition to this, in the majority of ionic liquids, the migration of protons takes place through the vehicular mechanism. The best part of using ionic liquid is that they can be used at a temperature above 100°C and preferably under anhydrous condition since the proton transport is independent of the water content of the electrolyte. The preparation of an ionic liquid is based on acid-base neutralization reaction via the transfer of a proton from an acid to a base. Protonic ionic liquids have been studied by several research groups from time to time.

Noda et al. have developed a Bronsted acid-base based ionic liquid using solid imidazole and bis(trifluoromethanesulfonyl) imide combination [29]. The above combination at 1:1 molar ratio forms a neutral salt that is stable up to 300°C. Here, the solid imidazole plays the role of a proton donor and acceptor, thereby helps in the improvement of oxygen reduction and hydrogen oxidation reactions. Ionic liquids are promising proton conducting materials for fuel cell devices. They provide excellent conductivity under anhydrous condition due to the water independent proton conduction process, however, the research in this field is at an early stage and a massive growth is expected.

4.5.2 Membranes based on polymer-ionic liquid

Polymeric membranes are the most preferable form of solid electrolytes since most of the electronic devices need a film-like material. The most common way of preparation is by physically mixing the ionic liquids with polymer matrix. Another preferred way of preparing ionic liquid-containing polymeric membranes is by free radical polymerization. Most of the ionic liquids are soluble in common vinyl monomers and hence can be easily polymerized by free radical route. A highly conductive ionic liquid has been synthesized by polymerizing methylmethacrylate in the presence of 1-ethyl-3-methyl imidazolium bis(trifluoromethanesulfone)imide. Using a low dose of crosslinker, a free-standing, flexible, and highly conductive film has been obtained. The membranes attain proton conductivity as high as 10^{-2} S cm^{-1} at room temperature and high ionic liquid content [30]. In another work, Langevin et al. have developed a proton-conducting membrane for high-temperature fuel cell application. The composite membranes are developed by the impregnation of triethylammonium trifluoromethanesulfonate in porous matrix material. The composite membranes are characterized by a storage modulus of 200 MPa and a proton conductivity of 2×10^{-2} S cm^{-1} at 130°C [31].

Solution casting is a common way of fabrication of ionic liquid-doped composite polymer membranes. Liew and co-workers have worked on the development of a proton conductive membrane based on PVA, 1-butyl-3-methyl-imidazolium chloride (BmImCl), and ammonium acetate. PVA is serving the role of the matrix material whereas ammonium acetate and BmImCl are typically playing the role of plasticizer and proton conductor, respectively. As the concentration of the ionic liquid increases, proton conductivity is gradually increased by the vehicular mechanism. At 50% BmImCl loading, the composite membrane exhibits proton conductivity as high as 5.74×10^{-3} S cm^{-1} and a maximum power density of 18 mW cm^{-2} at room temperature [32]. SPEEK is one of the widely explored polymer matrixes as membrane material. A unique combination of excellent thermal and chemical stability along with tunable proton conductivity and mechanical properties makes them a material of choice. In a recent study, Jothi et al. have reported a composite membrane based on SPEEK and alkylimidazolium phosphate ionic liquid by solution casting approach [33]. This class of ionic liquid has shown hydrolytic stability and proton conductivity under anhydrous conditions. The composite membrane has exhibited conductivity as high as 3×10^{-3} S cm^{-1} at 145°C under the dry condition with 50 wt.% ionic liquid content. The membranes are also characterized by improved thermal stability as compared to virgin membrane. These membranes have also attained a power density of 203 mW cm^{-2} and OCV of 0.83 V at 145°C. Other than SPEEK, sulfonated polyimides are another class of polymers used extensively for the purpose. Good thermal stability and film-forming ability establish them as a suitable material for fuel cell application.

Ionic liquid doped sulfonated polyimide composite membranes are studied further for anhydrous fuel cells [34, 35]. These membranes can be suitable for hydrogen fuel cells due to the excellent thermal stability, proton conductivity (10^{-2} S cm^{-1} at 120°C), current density (250 mA cm^{-2}), peak power density (mW cm^{-2}), and comparable gas permeation ability as that of commercial Nafion membranes.

4.5.3 Membranes based on polymerized ionic liquid

In the third approach, introduction of ionic liquid in the polymer matrix is done by polymerizing ionic liquid having vinyl groups. Polymerization can be carried out by using two basic strategies; first, the chemical modification of the existing polymer and the other is by directly polymerizing ionic liquid monomers via free radical route. Polymeric ionic liquids may have different types of structures, such as polyionics, including polycations and polyanions, polyzwitterions, and copolymers [36]. Ionic liquid monomers most commonly polymerized by the free radical route are *N*-vinyl imidazolium and methacryloyl styrenic groups. The photopolymerization of ionic liquid monomers is another effective way of incorporating ionic liquid into the polymer matrix. In different studies, researches have been conducted to synthesize polymeric ionic liquids by photopolymerization of imidazolium monomers [37–39]. Polymeric ionic liquids are also developed by the chemical modification route. The ionic liquids prepared via chemical modification are subjected to variable structural and polymerization degrees. The most common routes are by reactions of haloalkanes with imidazole containing polymers and by grafting of alkyl imidazole on to the haloalkyl-containing polymer repeating units [40]. However, polymerization of ionic liquid monomers may lower the proton conductivity due to the reduction in availability and mobility of the ions which in turn enhances the glass transition temperature. The only way out to suppress these effects is either by enhancing the ionic mobility through structural modification of the electrolyte or by increasing the number of mobile ions or charge carriers.

4.6 METAL-ORGANIC FRAMEWORKS (MOF)/COVALENT ORGANIC FRAMEWORKS (COF)-BASED ANHYDROUS PEM

MOF and COF are the new classes of material that can be designed with specific structural functionalities with tunable surface area and can be effectively utilized in water-assisted and anhydrous PEMs [41, 42]. The crystalline materials possess likely H-bonded conduction path as evidenced in X-ray analysis, owing to the specific molecular arrangements of ligands generates dynamic properties and proton conductivity in it. The crystalline porous material is made of metal ions coordinated with organic linkages giving rise to high surface area, structural variability and surface functionalization [43, 44]. The MOFs/coordination polymers were first reported by Kanda et al., however, the stability of these materials limit their use for proton conduction membrane [45]. Later researcher found that the materials can exhibit good conductivity under highly humid condition at temperature of

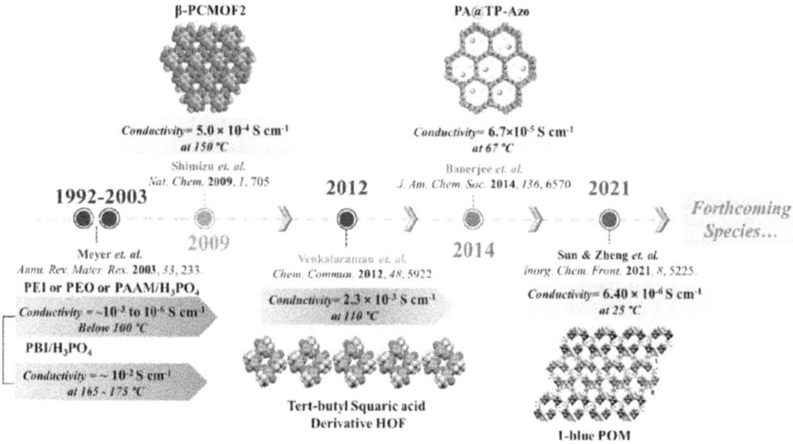

FIGURE 4.4 Examples of different proton conductors of crystalline platforms, i.e., MOF, COF, HOF, and POM [48] (Reprinted with permission)

about 60–80°C revealing their potential for low-temperature fuel cells [46]. Later on imidazole has been introduced in the nanochannel of MOF to improve proton conduction ability under anhydrous condition [47]. MOF-based proton conductors are explored significantly in the literature. They have been reported to contain carriers such as phosphates, organic heterocycles embedded in framework backbones to attain efficient packing of proton carrier with enhanced proton transport properties. COFs are crystalline material grabbed the attention of the researcher due to high surface area, porous architecture, strategic functionalization, high chemical and thermal properties. These are made of elements such as carbon, oxygen, sulfur, and hydrogen. Lighter elements result in low density and structural integrity by virtue of their unique architectural features and molecular arrangement. COFs are successfully explored in the field of sensors, photovoltaic, and electronics. COFs in addition to these features possess proton conductivity both in humid and non-humid condition and good chemical and thermal stability. Figure 4.4 depicts the examples of different proton conductors of crystalline platforms, i.e., MOF, COF, HOF, and POM, utilized in PEM.

Depending on the arrangements of the legends and the proton conductors, they can be classified as anhydrous MOF, anhydrous COFs, anhydrous hydrogen-bonded framework structures, and anhydrous polyoxometalates (POMs). The proton conduction behavior of these materials is mainly governed by the immobilization of charge carriers. The anhydrous proton conductivity may arise either intrinsically or extrinsically depending on the proton source, i.e., whether the proton source is inherently a part of the framework or externally loaded with some proton conductors to generate conduction [48]. The mechanism of conduction is mainly governed by hydrogen bond formation with the water molecules facilitating the efficient transport of the protons.

In a recent study, Divya et al. have reported incorporation of phosphorylated MOF/COF in SPEEK matrix and examined its suitability for high-temperature fuel cells [49]. The novel PEM is developed by incorporation of microporous MOF and

COF doped with phosphoric acid to impart high-temperature proton conductivity. The optimized membrane exhibits conductivity of 7.6×10^{-3} Scm^{-1} at 130°C under dehydrated state. The active functional groups present in the hybrid favor transport by acid-base interaction with the dopant phosphoric acid. In another study, Li et al. have prepared high-temperature PEMs containing multilayer structures by alternative deposition of carbon dot/MOF and SPEEK by spin coating [50]. The carbon dots/MOF composite is prepared by self-assembly technique and added in SPEEK. The carbon dot/MOF is again combined with phosphoric acid to further enhance the proton conductivity. The optimized composite membrane shows proton conductivity as high as $5.02 \pm 0.64 \times 10^{-2}$ S cm^{-1} at temperature of 160°C. Impressively, 1.53×10^{-2} Scm^{-1} conductivity is retained at temperature of 80°C even after 200 hours of continuous operation.

4.7 CHAPTER SUMMARY

PEMs are widely accepted as the promising materials for low-temperature fuel cells typically functioning in the temperature range 60–80°C in the presence of highly expensive platinum catalyst. The requirement of costly catalyst along with optimum hydration level to acquire usable proton conductivity limits the utilization of conventional PEMs in next-generation electric vehicles that somehow demand shift of research focus toward anhydrous PEM operable with cost-efficient oxide catalysts at a temperature of around 100–120°C. Triazole-containing PEM has been the recent interest of the researchers and has shown to be promising up to 120°C under low humidity condition. The anhydrous PEMs of different classes have been discussed in this chapter to give the readers an overview on the recent developments in this field for applications in small electronic devices operable at wide temperature range without requirement of complicated heat and water management.

REFERENCES

[1] Kalapos TL, Decker B, Every HA, Ghassemi H, Zawodzinski TA. Thermal studies of the state of water in proton conducting fuel cell membranes. *J Power Sources* 2007;172:14–9. https://doi.org/10.1016/j.jpowsour.2007.04.082.

[2] Schuster MFH, Meyer WH, Schuster M, Kreuer KD. Toward a new type of anhydrous organic proton conductor based on immobilized imidazole. *Chem Mater* 2004;16: 329–37. https://doi.org/10.1021/cm021298q.

[3] Smitha B, Sridhar S, Khan AA. Solid polymer electrolyte membranes for fuel cell applications—a review. *J Memb Sci* 2005;259:10–26. https://doi.org/10.1016/j.memsci.2005.01.035.

[4] Sahoo R, Mondal S, Pal SC, Mukherjee D, Das MC. Covalent-Organic Frameworks (COFs) as proton conductors. *Adv Energy Mater* 2021;11:2102300–33. https://doi.org/10.1002/aenm.202102300.

[5] Alaswad A, Omran A, Sodre JR, Wilberforce T, Pignatelli G, Dassisti M, et al. Technical and commercial challenges of Proton-Exchange Membrane (PEM) fuel cells. *Energies (Basel)* 2020;14:144–64. https://doi.org/10.3390/en14010144.

[6] Pal SC, Das MC. Superprotonic conductivity of MOFs and other crystalline platforms beyond 10^{-1} S cm^{-1}. *Adv Funct Mater* 2021;31:2101584–606. https://doi.org/10.1002/adfm.202101584.

[7] Bose S, Kuila T, Nguyen TXH, Kim NH, Lau K, Lee JH. Polymer membranes for high temperature proton exchange membrane fuel cell: Recent advances and challenges. *Prog Polym Sci* 2011;36:813–43. https://doi.org/10.1016/j.progpolymsci.2011.01.003.

[8] Pal SC, Mukherjee D, Oruganti Y, Lee BG, Lim D-W, Pramanik B, et al. Room-temperature superprotonic conductivity beyond 10^{-1} S cm^{-1} in a Co(II) coordination polymer. *J Am Chem Soc* 2024;146:14546–57. https://doi.org/10.1021/jacs.4c01113.

[9] Subianto S, Mistry MK, Choudhury NR, Dutta NK, Knott R. Composite polymer electrolyte containing ionic liquid and functionalized polyhedral oligomeric silsesquioxanes for anhydrous PEM applications. *ACS Appl Mater Interfaces* 2009;1:1173–82. https://doi.org/10.1021/am900020w.

[10] Yan F, Yu S, Zhang X, Qiu L, Chu F, You J, et al. Enhanced proton conduction in polymer electrolyte membranes as synthesized by polymerization of protic ionic liquid-based microemulsions. *Chem Mater* 2009;21:1480–4. https://doi.org/10.1021/cm900098r.

[11] Seel DC, Benicewicz BC, Xiao L, Schmidt TJ. *High-temperature polybenzimidazole-based membranes. Handbook of Fuel Cells – fundamentals, technology and applications.* 5th ed., John Wiley & Sons; 2009, p. 300–12.

[12] Özdemir Y, Özkan N, Devrim Y. Fabrication and characterization of cross-linked polybenzimidazole based membranes for high temperature PEM fuel cells. *Electrochim Acta* 2017;245:1–13. https://doi.org/10.1016/j.electacta.2017.05.111.

[13] Mader JA, Benicewicz BC. Sulfonated polybenzimidazoles for high temperature PEM fuel cells. *Macromolecules* 2010;43:6706–15. https://doi.org/10.1021/ma1009098.

[14] Kreuer KD. On the development of proton conducting materials for technological applications. *Solid State Ion* 1997;97:1–15. https://doi.org/10.1016/S0167-2738(97)00082-9.

[15] Kreuer KD. New proton conducting polymers for fuel cell applications. *Solid State Ionics: Sci & Technol* 1998:263–74.

[16] Münch W, Kreuer K-D, Silvestri W, Maier J, Seifert G. The diffusion mechanism of an excess proton in imidazole molecule chains: First results of an ab initio molecular dynamics study. *Solid State Ion* 2001;145:437–43. https://doi.org/10.1016/S0167-2738(01)00941-9.

[17] Zhou Z, Liu R, Wang J, Li S, Liu M, Brédas J-L. Intra- and intermolecular proton transfer in 1 *H* (2 *H*)-1,2,3-triazole based systems. *J Phys Chem A* 2006;110:2322–4. https://doi.org/10.1021/jp057265q.

[18] Zhou Z, Li S, Zhang Y, Liu M, Li W. Promotion of proton conduction in polymer electrolyte membranes by 1 *H* -1,2,3–triazole. *J Am Chem Soc* 2005;127:10824–5. https://doi.org/10.1021/ja052280u.

[19] Sevil F, Bozkurt A. Proton conducting polymer electrolytes on the basis of poly(vinylphosphonic acid) and imidazole. *J Phys Chem Solids* 2004;65:1659–62. https://doi.org/10.1016/j.jpcs.2004.04.001.

[20] Günday ST, Bozkurt A, Meyer WH, Wegner G. Effects of different acid functional groups on proton conductivity of polymer-1,2,4-triazole blends. *J Polym Sci B Polym Phys* 2006;44:3315–22. https://doi.org/10.1002/polb.20956.

[21] Erdemi H, Bozkurt A, Meyer WH. PAMPSA–IM based proton conducting polymer electrolytes. *Synth Met* 2004;143:133–8. https://doi.org/10.1016/j.synthmet.2003.10.022.

[22] Karadedeli B, Bozkurt A, Baykal A. Proton conduction in adipic acid/benzimidazole hybrid electrolytes. *Physica B Condens Matter* 2005;364:279–84. https://doi.org/10.1016/j.physb.2005.04.024.

[23] Yamada M, Honma I. Alginic acid–imidazole composite material as anhydrous proton conducting membrane. *Polymer (Guildf)* 2004;45:8349–54. https://doi.org/10.1016/j.polymer.2004.10.017.

[24] Lee YJ, Bingöl B, Murakhtina T, Sebastiani D, Meyer WH, Wegner G, et al. High-resolution solid-state NMR studies of poly(vinyl phosphonic acid) proton-conducting polymer: Molecular structure and proton dynamics. *J Phys Chem B* 2007;111:9711–21. https://doi.org/10.1021/jp072112j.

[25] Sevil F, Bozkurt A. Proton conduction in PVPA --benzimidazole hybrid electrolytes. *Turk J Chem* 2005;29:373–83.
[26] Yamada M, Honma I. Proton conducting acid–base mixed materials under water-free condition. *Electrochim Acta* 2003;48:2411–5. https://doi.org/10.1016/S0013-4686(03)00263-9.
[27] Göktepe F, Bozkurt A, Günday ŞT. Synthesis and proton conductivity of poly(styrene sulfonic acid)/heterocycle-based membranes. *Polym Int* 2008;57:133–8. https://doi.org/10.1002/pi.2335.
[28] Yu J, Yi B, Xing D, Liu F, Shao Z, Fu Y, et al. Degradation mechanism of polystyrene sulfonic acid membrane and application of its composite membranes in fuel cells. *Phys Chem Chem Phys* 2003;5:611–5. https://doi.org/10.1039/b209020a.
[29] Noda A, Susan Md ABH, Kudo K, Mitsushima S, Hayamizu K, Watanabe M. Brønsted acid–base ionic liquids as proton-conducting nonaqueous electrolytes. *J Phys Chem B* 2003;107:4024–33. https://doi.org/10.1021/jp022347p.
[30] Ueki T, Watanabe M. Macromolecules in ionic liquids: Progress, challenges, and opportunities. *Macromolecules* 2008;41:3739–49. https://doi.org/10.1021/ma800171k.
[31] Langevin D, Nguyen QT, Marais S, Karademir S, Sanchez J-Y, Iojoiu C, et al. High-temperature ionic-conducting material: Advanced structure and improved performance. *J Physi Chem C* 2013;117:15552–61. https://doi.org/10.1021/jp312575m.
[32] Liew C-W, Ramesh S, Arof AK. A novel approach on ionic liquid-based poly(vinyl alcohol) proton conductive polymer electrolytes for fuel cell applications. *Int J Hydrogen Energy* 2014;39:2917–28. https://doi.org/10.1016/j.ijhydene.2013.07.092.
[33] Jothi PR, Dharmalingam S. An efficient proton conducting electrolyte membrane for high temperature fuel cell in aqueous-free medium. *J Memb Sci* 2014;450:389–96. https://doi.org/10.1016/j.memsci.2013.09.034.
[34] Lee S-Y, Yasuda T, Watanabe M. Fabrication of protic ionic liquid/sulfonated polyimide composite membranes for non-humidified fuel cells. *J Power Sources* 2010;195:5909–14. https://doi.org/10.1016/j.jpowsour.2009.11.045.
[35] Lee S-Y, Ogawa A, Kanno M, Nakamoto H, Yasuda T, Watanabe M. Nonhumidified intermediate temperature fuel cells using protic ionic liquids. *J Am Chem Soc* 2010;132:9764–73. https://doi.org/10.1021/ja102367x.
[36] Ohno H. Design of Ion conductive polymers based on ionic liquids. *Macromol Symp* 2007;249–250:551–6. https://doi.org/10.1002/masy.200750435.
[37] Simons K, Nijmeijer K, Bara JE, Noble RD, Wessling M. How do polymerized room-temperature ionic liquid membranes plasticize during high pressure CO2 permeation? *J Memb Sci* 2010;360:202–9. https://doi.org/10.1016/j.memsci.2010.05.018.
[38] Bara JE, Hatakeyama ES, Gabriel CJ, Zeng X, Lessmann S, Gin DL, et al. Synthesis and light gas separations in cross-linked gemini room temperature ionic liquid polymer membranes. *J Memb Sci* 2008;316:186–91. https://doi.org/10.1016/j.memsci.2007.08.052.
[39] Bara JE, Gin DL, Noble RD. Effect of anion on gas separation performance of polymer–room-temperature ionic liquid composite membranes. *Ind Eng Chem Res* 2008;47:9919–24. https://doi.org/10.1021/ie801019x.
[40] Yuan J, Antonietti M. Poly(ionic liquid)s: Polymers expanding classical property profiles. *Polymer (Guildf)* 2011;52:1469–82. https://doi.org/10.1016/j.polymer.2011.01.043.
[41] Ogiwara N, Iwano T, Ito T, Uchida S. Proton conduction in ionic crystals based on polyoxometalates. *Coord Chem Rev* 2022;462:214524. https://doi.org/10.1016/j.ccr.2022.214524.
[42] Sun X, Liu S, Zhang S, Dang T, Tian H, Lu Y, et al. High proton conductivity achieved by the self-assembly of POM-based acid–base adduct in SBA-15 over a wide range from −40 to 85°C. *ACS Appl Energy Mater* 2020;3:1242–8. https://doi.org/10.1021/acsaem.9b02381.
[43] Zhou H-C "Joe", Kitagawa S. Metal–Organic Frameworks (MOFs). *Chem Soc Rev* 2014;43:5415–8. https://doi.org/10.1039/C4CS90059F.

[44] Zhou H-C, Long JR, Yaghi OM. Introduction to metal–organic frameworks. *Chem Rev* 2012;112:673–4. https://doi.org/10.1021/cr300014x.
[45] Kanda S, Yamashita K, Ohkawa K. A proton conductive coordination polymer. I.[N, N'-bis (2-hydroxyethyl)-dithiooxamido] copper (II). *Bull Chem Soc Jpn* 1979: 3296–301.
[46] Hurd JA, Vaidhyanathan R, Thangadurai V, Ratcliffe CI, Moudrakovski IL, Shimizu GKH. Anhydrous proton conduction at 150°C in a crystalline metal–organic framework. *Nat Chem* 2009;1:705–10. https://doi.org/10.1038/nchem.402.
[47] Bureekaew S, Horike S, Higuchi M, Mizuno M, Kawamura T, Tanaka D, et al. One-dimensional imidazole aggregate in aluminium porous coordination polymers with high proton conductivity. *Nat Mater* 2009;8:831–6. https://doi.org/10.1038/nmat2526.
[48] Mukherjee D, Saha A, Moni S, Volkmer D, Das MC. Anhydrous solid-state proton conduction in crystalline MOFs, COFs, HOFs, and POMs. *J Am Chem Soc* 2025;147: 5515–53. https://doi.org/10.1021/jacs.4c14029.
[49] Divya K, Liu Q, Murali R, Asghar MR, Liu H, Zhang W, et al. Integration of phosphorylated MOF/COF core-shell structures into hybrid membrane for high-temperature fuel cell application. *Appl Surf Sci* 2025;681:161544. https://doi.org/10.1016/j.apsusc.2024.161544.
[50] Li Q, Gao W, Zhang N, Gao X, Wu D, Che Q. Preparation of high temperature proton exchange membranes with multilayered structures through alternate deposition of carbon dots@Metal organic framework and Sulfonated Poly(Ether Ketone). *J Memb Sci* 2025;713:123306. https://doi.org/10.1016/j.memsci.2024.123306.

5 Composite polymer electrolyte membranes

5.1 NANOCOMPOSITE AND COMPOSITE PEM

Non-perfluorinated and hydrocarbon-based membranes face challenges of poor mechanical integrity and low proton conductivity as compared to the commercial membranes. To resolve these issues, non-perfluorinated and hydrocarbon-based membranes are often functionalized with hydrophilic functional groups such as sulfonic, carboxylic, and phosphoric acid. However, to achieve desired proton conductivity, the degree of functionalization needs to be sufficiently high. At high degree of functionalization, membranes suffer from excessive swelling which in turn reduces the mechanical strength of membranes. The addition of various nanomaterials is adopted to increase the proton conductivity of the PEM giving rise to the fabrication of composite PEM with tuneable properties. The addition of nanofillers may serve various purposes. Depending on the types of nanofillers, improvement can be observed in proton conductivity, thermal and mechanical stability, and oxidative stability. The addition of modifiers also helps to enhance the water retention capacity with reduction in fuel crossover leading to the enhancement in the fuel cell efficiency. Again, apart from types of nanofillers, its content, composition, distribution inside the polymer matrix will also play a significant role in deciding the properties of the composite PEM. Functionalization of the nanofillers is also adopted to obtain homogeneous distribution of the nanofillers inside the polymer matrix. In this chapter, we will briefly discuss about the different types of nanofillers that are incorporated to fabricate composite PEM, recent development in the field of nanocomposite and composite PEM, and future research trend to overcome the challenges encountered during the fabrication of composite PEM.

5.2 PREPARATION OF COMPOSITE POLYMER ELECTROLYTE MEMBRANES

Polymer nanocomposite-based polymer electrolyte membranes are prepared by two common methods, *in situ* and ex situ techniques. In in situ techniques, the filler particles or their precursor solutions are first dispersed thoroughly in the monomer solution followed by polymerization of the monomer, whereas in the ex situ technique, filler particles or their precursors are directly dispersed into the polymer matrix. In the ex situ technique, solution blending and melt mixing are the two most common techniques. Although the melt blending process is the most environmentally friendly process, however, mixing of the filler via melt mixing leads to poor membrane

DOI: 10.1201/9781003733744-5

properties due to insufficient dispersion and distribution of the fillers within the polymer matrix, partial degradation during heat-assisted mixing, etc. Hence, solution blending is accepted widely as the efficient technique for proper filler dispersion within the polymer matrix. Nanocomposite preparation by solution blending technique allows proper dispersion of the filler within the polymer matrix without decomposition of the host polymer matrix. The nanocomposite-based membranes are finally prepared by casting and drying. Again, the nature of the solvent also plays a crucial role in determining the membrane properties. Various organic solvents, such as DMAc, DMSO, and DMF, are used for the preparation of membranes. Among various organic solvents used for the preparation of membranes, DMAc is found to be the most effective one as per the earlier study [1]. Successful preparation of the nanocomposite PEM involves the proper selection of preparation method. The difference in the chemical nature of two components such as matrix polymer and filler phase always tends to form heterogeneous morphology. Hence, proper distribution

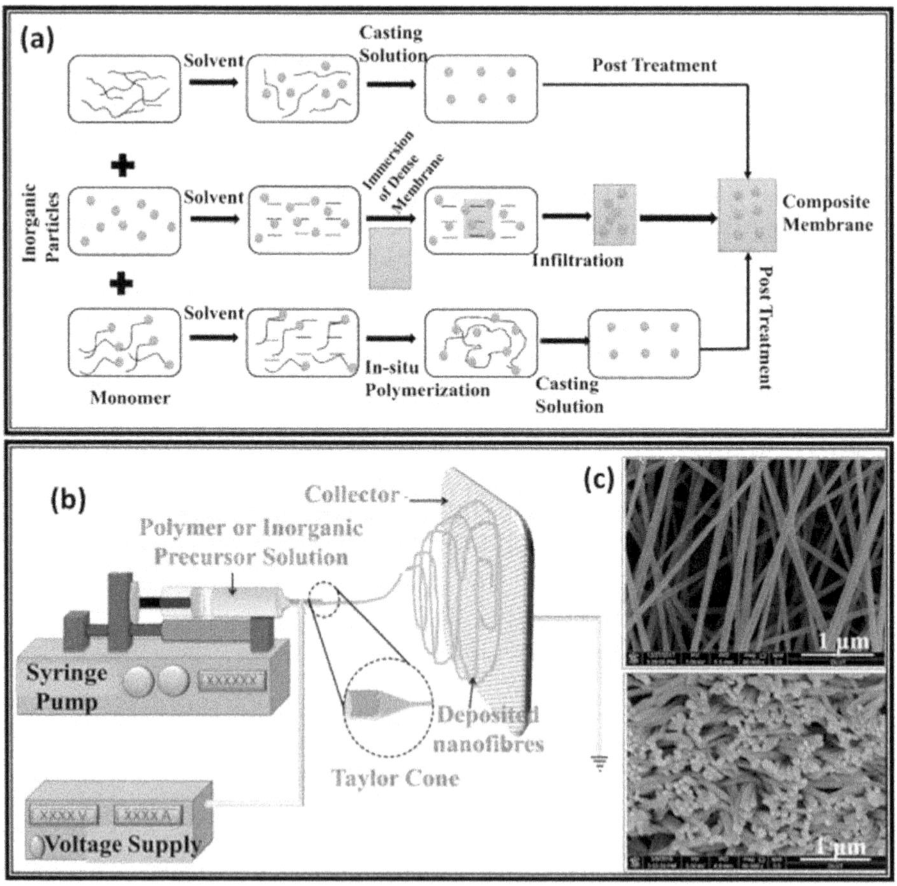

FIGURE 5.1 (a) Preparation of composite PEM, (b) electrospinning of polymer solution [2] and (c) SEM images of uniaxial and coaxial SPEEK mat [3] (Reprinted with Permission)

of the fillers within the polymer matrix is the foremost necessity to achieve homogeneous properties throughout the membrane. Different strategies have been taken to prevent the agglomeration of fillers within the polymer matrix, such as infiltration, recasting, and electrospinning. Figure 5.1 shows the various way of mixing the fillers with the polymers.

5.2.1 Infiltration

Infiltration is a common, simple, and low-cost method for the preparation of organic-inorganic nanocomposite membrane. In this method, a dense ionomer membrane is first prepared. The inorganic particles are dispersed in suitable solvent, and then dense ionomeric membrane is introduced into it. A suitable ionomeric membrane must possess solvated ionomeric channels or ionic clusters to allow the infiltration of inorganic particles of the same size within it. After the completion of the infiltration procedure, inorganic particles are stabilized via heat treatment or thermal annealing [4–6]. One of the major disadvantages of the infiltration process is the leaching of inorganic particles due to insufficient bonding with the polymer matrix. To overcome the leaching of inorganic fillers, infiltration is often combined with the polycondensation reaction to impart a strong bond between the filler and polymer matrix [7, 8]. Other methods such as electrostatic interaction and *in situ* polymerizations are also used to prevent the leaching of inorganic particles [9, 10]. The major challenge that lies in this method is the proper distribution of ionic channels throughout the membrane.

5.2.2 Recasting or blending

Recasting or blending is another common method of preparation of the nanocomposite membrane. The primary requirement is the proper dispersion of inorganic particles into the proposed solvent. The combination of blending and sol-gel technique allows a homogeneous distribution of the inorganic fillers not only in the hydrophilic part of the polymer backbone but also within the hydrophobic moiety which in turn leads to the development of a mechanically and thermally stable nanocomposite PEM [11]. Hence, recasting is considered to be a powerful technique for homogeneous distribution of the inorganic particles within the organic polymer matrix at an optimum filler loading. Di Vona et al. have developed titania-based SPEEK nanocomposite membranes by utilizing the *in situ* sol-gel technique [12]. Pan et al. have also used the combination of recasting and sol-gel technique to incorporate zirconia particles in the Nafion matrix via *in situ* hydrolysis of tetra butylzirconate [13]. Chen et al. have utilized ultrasonication to disperse sulfonated silica nanoparticles within the PBI matrix in DMAc medium. The nanocomposite membranes are observed to show improved interfacial compatibility between the components of the heterogeneous system [14]. Researchers have also tried ball milling to achieve enhanced filler distribution within the matrix. Krishnan et al. have studied the effect of ball milling speed and time on the dispersion of inorganic fillers in the SPES matrix. The study reveals that sufficient milling time and high milling speed lead to the formation of highly transparent nanocomposite PEM [15].

The main challenge encountered in the fabrication of composite PEMs by blending approach is the agglomeration or aggregation of the nanofillers leading to inhomogeneity in the overall properties of the composite PEM. Moreover, such inhomogeneous distribution of nanofillers is responsible for discontinuous proton conduction pathways restricting efficient conduction of protons in the polymer matrix. Modification of surface of the nanofiller is a common approach to address this issue and also to improve the compatibility of nanofilllers inside the polymer matrix.

5.2.3 Electrospinning

Electrospinning produces three-dimensional networks of electrospun nanofibers by a simple cost-efficient route. Various polymers are utilized as ionomeric polymeric matrices for fuel cells, e.g., Nafion, SPEEK, chitosan, SPEAS [16–20]. Nafion nanofibers are fabricated by several research groups in the presence of a carrier polymer to facilitate the electro-spinning process. However, the presence of non-ionic conducting polymer such as PEO reduces the effective concentration of acidic moieties and hence a decrease in proton conductivity has been evidenced. However, the electrospun nanofibers exhibit superior *in-plane* proton conductivity compared to the untreated one [21–23]. In a recent study, Klose et al. have developed sulfonated poly(ether ketone) reinforced Nafion composite membranes for fuel cells [24]. The reinforcement with proton-conducting sulfonated poly(ether ketone) produces highly durable fuel cells with good power density. The introduction of sulfonated poly(ether ketone) has improved the power density as compared to the reference due to a reduction in membrane resistance. Also, the membrane exhibits good chemical and long-term stability without promoting degradation. Proton conductivity of the existing membranes has also been improved by introducing sulfonated poly(aryl ether ketone) nanofibers. Li et al. have studied SPEEK based polymer electrolyte membranes [25]. The matrix polymers are fabricated by electrospinning and proton conductivity has been increased up to 0.37 S cm^{-1} compared to the conventional SPEEK membranes. Therefore, electrospinning technique has been successful in improving the conductivity by 37 times as compared to the solution casted membranes. Sulfonated carbon nanofibers are synthesized by electrospinning technique for the preparation of SPEEK composite membranes [26]. Incorporation of sulfonated carbon nanofibers of 100 nm diameter has effectively modified the proton transport channels of PEM and improved the methanol permeation resistance of polymeric membranes.

Membranes obtained by electrospinning are reported to be highly proton conductive as compared to the solution cast one. A new path of research has been opened for Nafion-based PEM via the development of electrospun Nafion nanofibers-based membranes. Dong et al. have demonstrated Nafion nanofibers with proton conductivity as high as 1.5 S cm^{-1} which is significantly higher in magnitude as compared to the bulk Nafion membrane. They used 0.1% of high molecular weight polymer (polyethylene oxide) as the carrier. This study also revealed that a reduction in fiber diameter is advantageous [27]. Li et al. have studied sulfonated polyether ether ketone ketone (SPEEKK) nanofibers-based PEM and investigated the effect of solution casted membrane and electrospun fiber based membrane on the fuel cell properties [25]. The electrospun fiber-based composite membrane has evidenced phase-separated morphology as compared to the solution-casted membrane and hence 37 times higher proton conductivity has been achieved. Sulfonated poly(phthalazinone ether sulfone ketone) (SPPESK) membranes are also prepared with the aim to improve through-plane proton conductivity via the electrospinning process [28]. In the presence of strong electric field, sulfonic acid groups try to aggregate on the surface of electrospun nanofibers resulting in very highly oriented through-plane proton conductivity in the nanofibers. In another work, Laudenslager et al. have reported about 10-fold improvement in proton conductivity as compared to the cast membranes [29]. Dong et al. have suggested that shear force is generated under a strong electric field and hence elongated and oriented ionic groups are observed along the fiber axis as evidenced in X-ray scattering [27]. In another study, Tamura and Kawakami have prepared composite membranes using sulfonated polyimide and nanofibers of sulfonated polyimide for PEMFC application. The polyimides are found to be oriented within the nanofibers. The membrane has exhibited improved proton conductivity as well as better methanol selectivity. Again the study reports that the proton conductivity value of the membranes is higher in the parallel direction as compared to the perpendicular one [30].

5.2.4 Activation of Membrane

To evaluate the membrane performance for fuel cell applications, it is essential to activate the membranes by chemical treatment. Chemical treatment of the membranes allows the removal of impurities and activation of the side chain functional groups such as sulfonic acid groups in the case of sulfonated membranes. Hudiono et al. have studied the effect of membrane activation on the proton conductivity and methanol permeation of porous oxide-based modified Nafion nanocomposite membranes [31]. The membranes are fully activated by subsequent treatment of H_2O_2, H_2O, H_2SO_4, H_2O, and in the later stage by thermal annealing. Properties of the membranes are also compared with acid-activated and water-activated membranes. The process of activation has a profound effect on the methanol permeation and proton conductivity of the membranes.

5.3 MECHANISM OF TRANSPORT IN NANOCOMPOSITE PEM

The transport of protons in composite PEM is primarily dependent on the types of nanofillers and polymer matrices. In PEM, majority of the polymer matrices contains sulfonic acid or any other hydrophilic functionality leading to the creation of a proton conduction channel through which the transport phenomenon occurs. The addition of nanofillers containing conducting proton (H^+ ions) provides additional protons in the system. Some water retentive fillers holds water in its crystal structure such as heteropolyacids which again favors conduction process by vehicle mechanism. In any case, the presence of nanofillers or nanofibers as the modifier plays a crucial role in determining the overall performance of the composite PEM. The transport mechanism observed in composite PEM fabricated with functionalized and non-functionalized modifier is schematically represented in Figure 5.2. The aggregation of fillers is observed in places for non-functionalized PEM disrupting the continuous path for proton transport

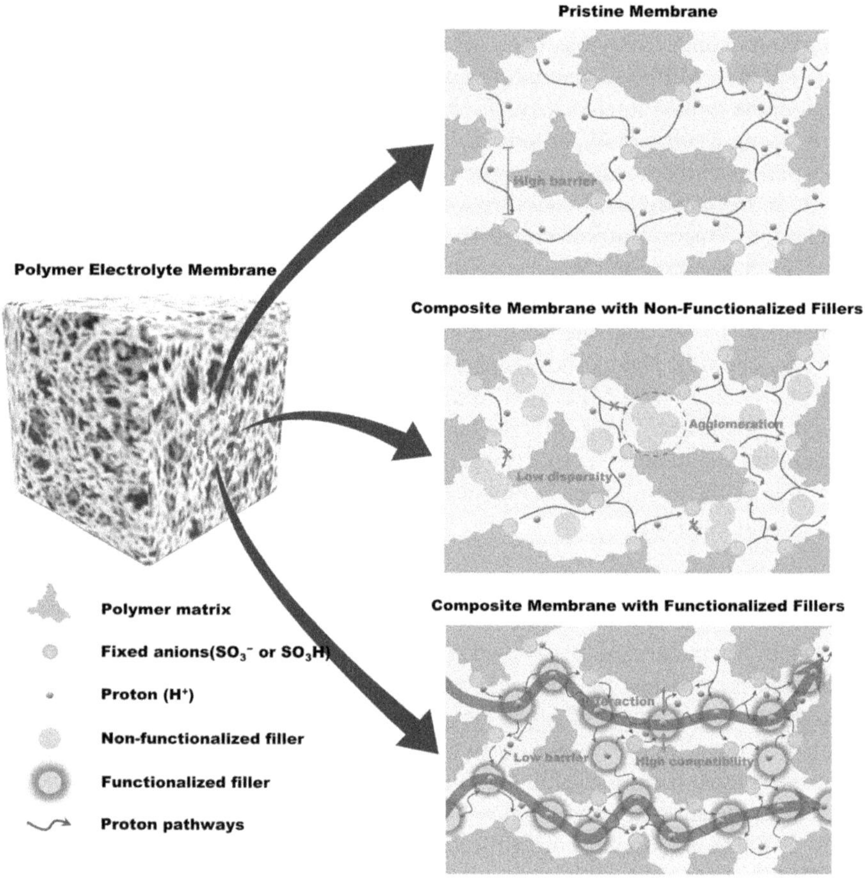

FIGURE 5.2 Mechanism of proton conduction in functionalized and non-functionalized composite PEM fabricated with functionalized and non-functionalized nanofillers [33] (Reprinted with Permission)

whereas in functionalized PEM, the modifier is distributed uniformly and hence able to maintain a continuous channel for easier and convenient proton transport. The functionalization also provides compatibility between the fillers and polymer. Study reveals that introduction of sulfonic acid groups in carbon nanotube improves interfacial interaction with the SPEEK composite and the conductivity and mechanical strength have been improved as compared to pure SPEEK matrix [32].

In another study, Steffy et al. prepared a composite PEM with Nafion and sulfonated carbon nanotube (CNT) by grafting sulfonic acid on the surface of the CNT [34]. The composite membrane filled with 0.5 wt.% functionalized multiwalled carbon nanotube (MWCNT) loading exhibited 11-fold higher proton conductivity as compared to the Nafion. As the loading is increased to 0.5 wt.%, the uptake and proton conductivity decline due to possible aggregation of MWCNT inside the matrix. Hence, an optimized amount of MWCNT is essential to obtain maximum benefit of nanofillers.

5.4 SIGNIFICANCE OF MODIFIERS UTILIZED IN COMPOSITE PEM

Nanofillers and micron-sized particles are incorporated inside the polymer matrix to fabricate nanocomposite and polymer nanocomposite membranes for PEMFC application. The inorganic nanoparticles are of two types, porous or non-porous fillers. SiO_2 and TiO_2 are the common non-porous fillers utilized for the fabrication of composite membranes for PEMFC devices [35]. Zeolites, however, are porous in nature. In PEM, the fillers are introduced with various purposes, mainly to increase the proton conductivity and water retention ability. In addition to this, oxides and carbon nanomaterials are also beneficial to enhance the thermal and mechanical stability of the polymer matrices as well as fuel permeability. The addition of proton conductive fillers directly influences the performance by providing more proton conduction sites whereas water retentive fillers help to retain water inside their porous structures. The water present inside the porous structure improves the proton conductivity where the conduction phenomenon is governed by water assisted *Vehicular* mechanism.

Silica and titania are the most commonly used solid non porous fillers [36, 37]. Others are clay, graphene, heavy metals, etc. Incorporation of these inorganic materials inside the polymer matrix, leads to the enhancement in thermal and mechanical properties which is an additional benefit for the construction of robust MEA. Mesoporous silica is another class of silica that is beneficial to be utilized as modifier in PEM. Inside the mesoporous structure, it retains substantial amount of water that facilitate proton conduction. The presence of silica also offers resistance to permeation that makes its more suitable for direct methanol fuel cells (DMFCs).

Clay is water retentive filler commonly utilized in PEMFCs [35, 38]. Clay possesses voids and pores that help in the retention of the water. Cationic clay, e.g., montmorillonite (MMT) belongs to 2:1 phyllosilicate family. MMT is composed by two tetrahedral SiO_2, one octahedral magnesium hydroxide or aluminium hydroxide in its structure. Modified organoclay, also known as cloisite is commonly added in SPEEK based hydrocarbon type matrices to fabricate composite membranes with good proton conductivity and mechanical strength [39]. The plate like structure in clay also provides additional benefit of barrier properties help in the reduction of fuel cross over in the device. The water uptake properties of the PEM can be enhanced by the addition of hydrophilic metal oxides, such as zeolite, silica, zirconia, which, in turn, gives a positive effect on the proton conductivity of PEM. Figure 5.3 displays the proton conductivity and polarization curves for Nafion and Nafion-based composite membranes fabricated with SiO_2, TiO_2, WO_3, PWA, etc., at 100°C temperature and 100% RH. At low RH, the SiO_2-filled composite membranes exhibit better proton conductivity as compared to the Nafion 115 membranes.

Proton conductivity of the composite membrane is further improved by addition of solid conducting fillers, such as heteropolyacids, e.g., phosphotungstic acid and silicotungstic acid [41]. The solid conducting fillers have potential to improve the proton conductivity by providing additional H^+ ions in the medium that facilitates the proton transport inside the membrane. The solid conducting filler increases the conductivity however, the water soluble hydrophilic filler has the tendency to be leach out from the composite membrane during prolonged operation. Researcher worked on controlling the leaching tendency of these fillers to improve

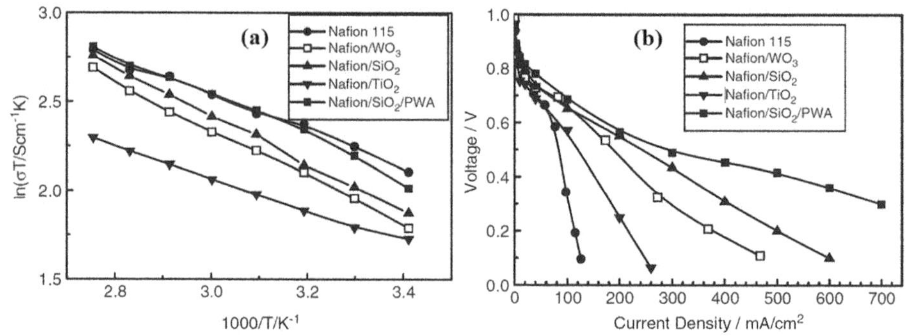

FIGURE 5.3 (a) Proton conductivity of Nafion and composite membranes, (b) Polarization curves for PEMFC device fabricated with Nafion and Nafion composite membranes at 100°C and 70% RH. Reprinted with Permission from [40].

the durability of composite membranes [42]. Heteropolyacids provide good acidity with high conductivity. Moreover, the water molecule present inside the crystal structure of heteropolyacids also helps in the water assisted proton conduction mechanism inside the membrane. In addition to good conductivity in hydrated state, heteropolyacids also exhibit good conductivity under dehumidified condition and hence, the decline in the proton conductivity with prolonged operation of fuel cell can also be managed to some extent in these composite PEMs. To control the leaching of these conducting fillers, the protons of tungstophosphoric acid is partially replaced with other metal ions such as cesium and aluminium in earlier study showing the evidence of long time retention of fillers inside the polymer matrix and steady maintenance of proton conductivity during operation of fuel cell devices [41, 43, 44].

Carbon nanomaterials such as carbon nanotubes and graphene are utilized as the additives in SPEEK membranes due to their ability to improve the mechanical properties and restricting the methanol permeation in DMFC. However, the contribution of carbon nanomaterials toward electrical conductivity has to be adjusted to achieve efficient proton conduction and thus the fuel cell performance in the device. Again, the carbon nanotubes also suffer from inherent problem of aggregation due to the high surface energy that restricts the dispersion of nanotubes inside the membrane. Uniform dispersion of nanotubes inside the membrane can be achieved by the incorporation of polar functionality within the nanofillers. Incorporation of sulfonated and carboxylated functional groups in the CNT improves the proton conductivity due to the insertion of proton conduction sites. The presence of hydrophilic groups in the nanotubes also helps in improving the compatibility of the nanofillers inside the polymer matrices leading to better dispersion and improved properties in the composites. Graphene is another carbon nanomaterial of interest which has attracted attention of the scientist and researchers in various fields due to excellent combination of properties. Graphene in oxide form, i.e., graphene oxide (GO) contains various polar groups, such as oxygen, hydroxyl, and carboxylic, that contribute to the interesting combination of properties [45, 46]. GO is amphiphilic

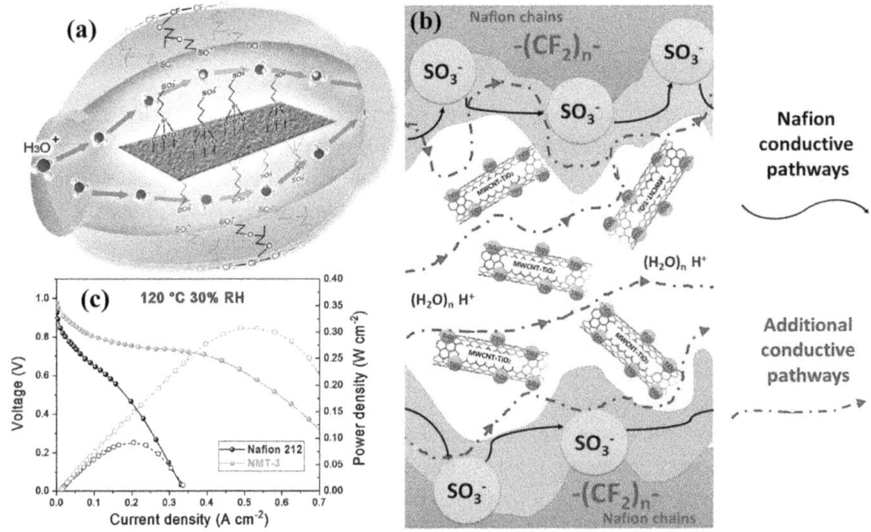

FIGURE 5.4 (a) Conduction mechanism in functionalized GO/Nafion composite PEM [50]; (b) Conduction pathways in MWCNT/Nafion composite PEM, and (c) Electrochemical polarization of MWCNT/Nafion composite PEM device [49] (Reprinted with Permission)

in nature and shows existence of hydrophilic and hydrophobic regions. GO exhibits poor electrical properties and good proton conduction ability due to the presence of hydrophilic groups on its surface [47]. GO also provides mechanical strength to the PEM due to strong covalent bonding. Experimental results show that graphene oxide paper can resist fuel crossover and hence restricting fuel permeability across the electrodes [48]. Figure 5.4 (a) displays the mechanism of proton conduction in the composite PEM fabricated with functionalized GO/ Nafion. The effect of MWCNT-TiO_2 loading (from 1 to 5 wt.%) on the fuel cell properties of Nafion 212 has been investigated in the recent study [49]. Figure 5.4 (b) and (c) depict the mechanism of conduction and polarization curves for the composite PEM. The electrochemical performance of the device is evaluated at 120°C temperature and 30% RH. The Nafion/MWCNT-TiO_2 composite fabricated at 3 wt.% filler loading (designated by NMT-x) is shown in Figure 5.4 (c).

Metal organic framework (MOF) has emerged as another porous nanofiller to be utilized as proton conductor for PEMFCs [51, 52]. MOFs are typically composed of metal ions and organic ligands bonded with coordination bond. MOFs are characterized by high surface area due to tunable porous architectures. They are utilized efficiently as the electrolyte material in batteries due to presence of redox metal centers. Zeolite, an aluminosilicate class of material, is also utilized extensively in the PEM [53–55]. The composite membranes with zeolite are able to hold water molecules in the porous crystal structures of zeolite which facilitates the conduction of protons inside the composite membrane. The addition of zeolite also improves other properties of interest such as mechanical and thermal stability as well as water retention ability.

5.5 DIFFERENT TYPES OF COMPOSITE PEM

5.5.1 Sulfonated polymer based nanocomposite PEM

Inorganic-organic nanocomposite polymer electrolytes have been extensively studied in the literature. Sulfonated nanocomposite PEM is comprehensively explored by various research groups over the years due to low cost, tunable properties and ease of synthesis of the sulfonated polymer. However, simultaneously achieving proper mechanical strength, thermal stability, and realistic proton conductivity always remain challenging. SPEEK-based composite membranes have been extensively utilized by the incorporation of various nanofillers, such as silica, titania, clay, MMT, zeolite, and heteropolyacids [12, 41, 56–58]. In our earlier chapter, we have extensively discussed the various aspects of SPEEK. Sulfonated polyether imide (SPI)-based PEM is also explored for low-temperature fuel cell devices due to superior oxidative resistance and thermal stability as well as low fuel permeability. These membranes also possess excellent mechanical properties along with proton conduction ability. However, the solubility of SPI membrane in water is a major challenge which leads to excessive swelling and decreased dimensional stability [59, 60]. Zhou et al. fabricated sulfonated CNT/SPEEK composite membranes for DMFC devices [61]. The inclusion of sulfonated CNT leads to enhancement in proton conductivity and tensile strength. Moreover, sulfonated CNT leads to reduced methanol permeability in the membrane which is essential for DMFC devices. In another study, polystyrene styrene acid (PSSA) modified MWCNT has been incorporated in SPEEK matrix. The nanocomposite membrane so fabricated, exhibited improved performance compared to SPEEK and commercial Nafion membrane. Organically modified clay has also been added in SPEEK matrix at a loading level of 3–5 wt.%. The composite membrane exhibited proton conductivity of 5.6×10^{-2} S cm^{-1} at 3 wt.% of filler loading in comparison to the proton conductivity of 2×10^{-3} and 1.6×10^{-2} S cm^{-1} for commercial Nafion 117 and synthesized SPEEK, respectively [62]. Organically reacted MMT has been added in the SPEEK matrix at 1 wt.% loading to improve the properties of SPEEK [63]. The performance and selectivity of the composite membranes were further enhanced by modifying with various nanofillers such as sulfanilic acid and propane sulfonic acid. The presence of hydrophilic groups boosted proton transport inside the membrane. Sulfonic acid modification has also been done on silica and then the sulfonated silica has been incorporated in SPEEK membrane for the fabrication of composite PEM at an optimum loading of 3 wt.% [64]. The membrane exhibited improved proton conductivity however the barrier properties was observed to be reduced due to excessive swelling. TiO_2 has been another common nanofiller incorporated into polymer matrices like SPEEK and PVA, both in functionalized and non-functionalized form by various researchers [65, 66]. Kalid et al. added TiO_2 in the PVA matrix and used maleic acid as crosslinking agent. The composite membrane at optimum loading exhibited improved thermal, mechanical, proton conduction properties that ensured its viability as PEM for low temperature fuel cell devices [67].

Apart from nanoparticles, nanofibers, e.g., carbon nanofibers (CNF) with tunable nanostructures have also been prepared by electrospinning technique and incorporated in polymer matrices for fabrication of composite PEM. Di et al. have studied continuous CNF (CCNF) mats as a filler to prepare composite PEM [68]. The results reveal that the incorporation of CCNF is effective in improving various fuel cell related properties such as water uptake, mechanical properties, and proton conductivity. However, a decrease in methanol selectivity has been evidenced. Liu et al. have prepared sulfonated CNF (SCNF)/SPEEK nanocomposite PEM for fuel cells [26]. CNF mat prepared by electrospinning technique has been subsequently sulfonated to obtain SCNF. Excellent bonding between the SCNF and SPEEK is obtained by hydrogen bonding among the sulfonic acid groups of SPEEK and SCNF in the presence of water molecules. The composite membranes exhibit good water uptake, electrochemical properties, and mechanical strength as compared to the neat SPEEK. However, a very high amount of SCNF loading in the polymer matrix may lead to aggregation of the filler and hence can reduce the proton conductivity of membranes. The best combination of properties is reported for the composite membranes having 1 wt.% SCNF which has shown proton conductivity comparable to Nafion 117 (0.128 S cm^{-1}) and substantially lower methanol permeability of 5.12×10^{-7} cm^2 s^{-1}. A good PEM should exhibit good proton conductivity along with methanol permeability. To achieve these properties, various electrospinning parameters such as concentration of solution, voltage, humidity, and distance from collector to spinnerate, are needed to be controlled. However, most common drawback of electrospun fibers is their low mechanical strength. Owing to their small diameter, fibers are very much prone to breakage [69]. Hence often a reinforcing material is incorporated to improve the mechanical strength as well as to withstand the effect of elevated temperature and pressure during the prolonged fuel cell operation.

5.5.2 Fluorinated polymer-based naonocomposite PEM

Combinations of fluorinated polymer Nafion and non-Nafion type membrane are the excellent choice for the fabrication of PEM for possible application in low temperature fuel cells due to robust structural integrity both in terms of mechanical and thermal stability. Composite membranes also provide good chemical resistance and proton conductivity due to the presence of hydrophilic moieties. The presence of highly electronegative fluorine atom in –C-F bonds leads to very high ionic bond strength in fluorinated polymers as compared to –C-H bond containing hydrocarbons. The polarization of electron clouds in –C-F bonds is responsible for high strength, chemical and thermal resistance in Nafion and other fluorinated polymers.

Nafion and other fluorinated polymers are extensively utilized for the fabrication of nanocomposite PEM in the literature. Sulfonated polytetrafluoro ethylene (PTFE) is one of the common choices due to the combination of properties such as thermal resistance, chemical resistance, and proton conductivity. Fluoropolymers-based composite membranes have been successfully prepared with various nanofillers such as inorganic oxides, carbon nanomaterials, and clay. The composite membranes possess excellent combination of properties along with efficient water management and fuel crossover [70]. Ionic liquid is doped in different fluoropolymers, e.g., poly(vinylidene-fluoride) and its copolymer with hexafluoropropylene, poly(vinylidene fluoride-co-trifluoroethylene) and its derivative and evaluated for crystallinity, electrochemical properties and battery performance. Composite membranes have been observed to show good proton conductivity as well as electrochemical performance [71]. PTFE has also used in combination with SiO_2 and HPO_3 for its viability in medium to high temperature fuel cells. The PTFE/SiO_2/HPO_3 composite membranes are investigated at 120° to 200°C. The composite membrane displays proton conductivity of about 0.14 S cm^{-1} at a very low RH of 0.5% [72]. In another work, fluorinated nanocomposite PEM is fabricated by infiltration of Nafion in PTFE fiber mats through dip coating process and investigated the fuel cell properties for PEMFC application [73]. The swelling of membrane has been controlled and through plane conductivity has been improved as compared to the Nafion 117 at 120°C showing its potential for fuel cell devices over 100°C. Aligned PTFE nanofibers are fabricated using electrospinning technique and used as reinforcement for fabrication of composite PEM with polyethylene oxide matrix. The electrospun nanofiber reinforced composite membranes exhibited good dimensional stability and proton conductivity [74]

5.5.3 Imidazole-based naonocomposite PEM

Functionalized hydrocarbon polymer-based membranes serve as excellent material for low-temperature fuel cell devices, however, the proton conduction mechanism is largely dependent on the presence of water due to water assisted vehicular mechanism. The conductivity of these membranes fall drastically with the progressive operation of device as the membrane gets dried. Medium and high temperature PEM mainly contains heteroaromatic polymers, e.g., pyrazole, triazole, and imidazole. They exhibit efficient proton conduction at high temperature for prolonged operation period [75]. Hydrocarbon polymers are often used in combination with heteroatom containing modifiers, e.g., ionic liquids and/or blended with heteroaromatic polymers containing imidazole-based groups to improve the proton conductivity [76, 77]. The N atoms present in –N-H groups help to increase acid doping and hence improve the proton conductivity. Sulfonated polyimides are also explored by the researcher as PEM for fuel cell devices due to outstanding proton conductivity along with thermal, mechanical, and structural integrity. In sulfonated polyimide, the hydrophilic sulfonated groups contribute to water-assisted conductivity, whereas the imide group provides additional proton conduction sides and supports to retain proton conductivity during prolonged operation. Sulfonated PBI membranes have exhibited good proton conductivity and structural integrity [78]. These membranes also support low fuel crossover and can manage water accumulation related problems effectively. However, complicated synthesis process makes it difficult to utilize at the commercial level. Among various non-fluorinated membranes, sulfonated polyether sulfones and sulfonated poly(arylene ether ketone)-based composite PEM possess good oxidative and hydrolytic stability. In another study, sulfonated polyimide/PBI blend and perovskite nanoparticle-based composite have been prepared by solution casting approach [79]. The composite membrane has been further doped with phosphoric acid to improve the conductivity. The nanocomposite membrane exhibited a favorable interaction between the sulfonic acid and triazole groups. The –N-H groups of benzimidazole interact with the oxygen vacancy present in the nanoparticle. The membrane has exhibited excellent proton conductivity of 131 mS cm^{-1} at 180°C temperature and at a very low relative humidity of about 8% indicating immense potential for high temperature fuel cell devices. Oxy-polybenzimidazole (OPBI)/ionic liquid doped silica composite membrane has been studied recently [80]. The composite membrane has showed improved properties due to favorable interaction between the matrix and the reinforcement at the interfaces through the formation of strong H-bonding. At 3 wt.% filler loading, the proton conductivity has been observed to reach 0.219 S cm^{-1} at 160°C temperature which is four times better as compared to OPBI base membrane.

5.6 CHAPTER SUMMARY

This chapter briefly discusses the emergence of polymer composite and nanocomposite PEM containing various nanofiller modifiers. The composite has been made with the aim to improve the device performance and durability. Modifiers can serve various purposes such as improving proton conductivity, water retention capability, thermal stability, mechanical strength, durability during prolonged operation leading to an efficient polymer electrolyte membrane that can serve for a long service period. The nanofillers can be of porous and non-porous types. The common nanofillers explored are silica, titania, zirconia, aluminosilicates, zeolites, clay, and carbon nanomaterials. The performance of the composite PEMs is dependent on the content, type, and distribution of the modifiers inside the polymer matrices. The dispersion of the modifiers inside the matrix is improved by proper functionalization of the nanofilers. The functionalization is introduced with the aim to provide hydrophilic domains in the composite in order to facilitate the water assisted proton conduction mechanism in composite PEM. The carbon and layered silicates are beneficial since these can improve the barrier properties of the matrix and helps in reduction of fuel crossover. In this chapter, we have also discussed different types of polymer matrices and different types of nanofillers that are incorporated to fabricate composite and nanocomposites for PEM fuel cell devices.

REFERENCES

[1] Kaliaguine S, Mikhailenko SD, Wang KP, Xing P, Robertson G, Guiver M. Properties of SPEEK based PEMs for fuel cell application. *Catal Today* 2003;82:213–22. https://doi.org/10.1016/S0920-5861(03)00235-9.

[2] Sood R, Cavaliere S, Jones DJ, Rozière J. Electrospun nanofibre composite polymer electrolyte fuel cell and electrolysis membranes. *Nano Energy* 2016;26:729–45. https://doi.org/10.1016/j.nanoen.2016.06.027.

[3] Yuan Q, Fu Z, Wang Y, Chen W, Wu X, Gong X, et al. Coaxial electrospun sulfonated poly (ether ether ketone) proton exchange membrane for conductivity-strength balance. *J Memb Sci* 2020;595:117516–25. https://doi.org/10.1016/j.memsci.2019.117516.

[4] Damay F. Transport properties of Nafion™ composite membranes for proton-exchange membranes fuel cells. *Solid State Ion* 2003;162–163:261–7. https://doi.org/10.1016/S0167-2738(03)00238-8.

[5] Deng Q, Moore RB, Mauritz KA. Novel nafion/ORMOSIL hybrids via in situ sol-gel reactions. 1. probe of ORMOSIL phase nanostructures by infrared spectroscopy. *Chem Mater* 1995;7:2259–68. https://doi.org/10.1021/cm00060a012.

[6] Formato RM, Robert FK, Paul O, Nelson L, Leslie SR. Composite solid polymer electrolyte membranes. US7052793B2, 2006.

[7] Hagihara H, Uchida H, Watanabe M. Preparation of highly dispersed SiO_2 and Pt particles in Nafion®112 for self-humidifying electrolyte membranes in fuel cells. *Electrochim Acta* 2006;51:3979–85. https://doi.org/10.1016/j.electacta.2005.11.012.
[8] Mauritz KA, Stefanithis ID, Davis S V., Scheetz RW, Pope RK, Wilkes GL, et al. Microstructural evolution of a silicon oxide phase in a perfluorosulfonic acid ionomer by an in situ sol–gel reaction. *J Appl Polym Sci* 1995;55:181–90. https://doi.org/10.1002/app.1995.070550120.
[9] Liu Y, Nguyen T, Kristian N, Yu Y, Wang X. Reinforced and self-humidifying composite membrane for fuel cell applications. *J Memb Sci* 2009;330:357–62. https://doi.org/10.1016/j.memsci.2009.01.011.
[10] Tripathi BP, Shahi VK. SPEEK–zirconium hydrogen phosphate composite membranes with low methanol permeability prepared by electro-migration and in situ precipitation. *J Colloid Interface Sci* 2007;316:612–21. https://doi.org/10.1016/j.jcis.2007.08.038.
[11] Xu T. Ion exchange membranes: State of their development and perspective. *J Memb Sci* 2005;263:1–29. https://doi.org/10.1016/j.memsci.2005.05.002.
[12] Divona M, Ahmed Z, Bellitto S, Lenci A, Traversa E, Licoccia S. SPEEK-TiO_2 nanocomposite hybrid proton conductive membranes via in situ mixed sol–gel process. *J Memb Sci* 2007;296:156–61. https://doi.org/10.1016/j.memsci.2007.03.037.
[13] Pan J, Zhang H, Chen W, Pan M. Nafion–zirconia nanocomposite membranes formed via in situ sol–gel process. *Int J Hydrogen Energy* 2010;35:2796–801. https://doi.org/10.1016/j.ijhydene.2009.05.110.
[14] Suryani, Chang Y-N, Lai J-Y, Liu Y-L. Polybenzimidazole (PBI)-functionalized silica nanoparticles modified PBI nanocomposite membranes for proton exchange membranes fuel cells. *J Memb Sci* 2012;403–404:1–7. https://doi.org/10.1016/j.memsci.2012.01.043.
[15] Krishnan NN, Henkensmeier D, Jang JH, Kim H-J, Rebbin V, Oh I-H, et al. Sulfonated poly(ether sulfone)-based silica nanocomposite membranes for high temperature polymer electrolyte fuel cell applications. *Int J Hydrogen Energy* 2011;36:7152–61. https://doi.org/10.1016/j.ijhydene.2011.03.015.
[16] Jonghyun C, Kyung Min L, Ryszard W, Peter NP, Patrick TM. Nanofiber network ion-exchange membranes. *Macromolecules* 2008;41:4569–72.
[17] Laforgue A, Robitaille L, Mokrini A, Ajji A. Fabrication and characterization of ionic conducting nanofibers. *Macromol Mater Eng* 2007;292:1229–36. https://doi.org/10.1002/mame.200700200.
[18] Lee C, Jo SM, Choi J, Baek K-Y, Truong YB, Kyratzis IL, et al. SiO_2/sulfonated poly ether ether ketone (SPEEK) composite nanofiber mat supported proton exchange membranes for fuel cells. *J Mater Sci* 2013;48:3665–71. https://doi.org/10.1007/s10853-013-7162-7.
[19] Ojha SS, Stevens DR, Hoffman TJ, Stano K, Klossner R, Scott MC, et al. Fabrication and characterization of electrospun chitosan nanofibers formed via templating with polyethylene oxide. *Biomacromolecules* 2008;9:2523–9. https://doi.org/10.1021/bm800551q.
[20] Sanders EH, McGrady KA, Wnek GE, Edmondson CA, Mueller JM, Fontanella JJ, et al. Characterization of electrosprayed Nafion films. *J Power Sources* 2004;129:55–61. https://doi.org/10.1016/j.jpowsour.2003.11.020.
[21] Choi SW, Fu Y-Z, Ahn YR, Jo SM, Manthiram A. Nafion-impregnated electrospun polyvinylidene fluoride composite membranes for direct methanol fuel cells. *J Power Sources* 2008;180:167–71. https://doi.org/10.1016/j.jpowsour.2008.02.042.
[22] Ballengee JB, Pintauro PN. Morphological control of electrospun nafion nanofiber mats. *J Electrochem Soc* 2011;158:B568–72. https://doi.org/10.1149/1.3561645.
[23] Bajon R, Balaji S, Guo SM. Electrospun nafion nanofiber for proton exchange membrane fuel cell application. *J Fuel Cell Sci Technol* 2009;6:031004–9. https://doi.org/10.1115/1.3005577.

[24] Klose C, Breitwieser M, Vierrath S, Klingele M, Cho H, Büchler A, et al. Electrospun sulfonated poly(ether ketone) nanofibers as proton conductive reinforcement for durable Nafion composite membranes. *J Power Sources* 2017;361:237–42. https://doi.org/10.1016/j.jpowsour.2017.06.080.
[25] Li X, Hao X, Xu D, Zhang G, Zhong S, Na H, et al. Fabrication of sulfonated poly(ether ether ketone ketone) membranes with high proton conductivity. *J Memb Sci* 2006;281: 1–6. https://doi.org/10.1016/j.memsci.2006.06.002.
[26] Liu X, Yang Z, Zhang Y, Li C, Dong J, Liu Y, et al. Electrospun multifunctional sulfonated carbon nanofibers for design and fabrication of SPEEK composite proton exchange membranes for direct methanol fuel cell application. *Int J Hydrogen Energy* 2017;42:10275–84. https://doi.org/10.1016/j.ijhydene.2017.02.128.
[27] Dong B, Gwee L, Salas-de la Cruz D, Winey KI, Elabd YA. Super proton conductive high-purity nafion nanofibers. *Nano Lett* 2010;10:3785–90. https://doi.org/10.1021/nl102581w.
[28] Gong X, He G, Wu Y, Zhang S, Chen B, Dai Y, et al. Aligned electrospun nanofibers as proton conductive channels through thickness of sulfonated poly (phthalazinone ether sulfone ketone) proton exchange membranes. *J Power Sources* 2017;358:134–41. https://doi.org/10.1016/j.jpowsour.2017.05.022.
[29] Laudenslager MJ, Scheffler RH, Sigmund WM. Electrospun materials for energy harvesting, conversion, and storage: A review. *Pure Appl Chem* 2010;82:2137–56. https://doi.org/10.1351/PAC-CON-09-11-49.
[30] Tamura T, Kawakami H. Aligned electrospun nanofiber composite membranes for fuel cell electrolytes. *Nano Lett* 2010;10:1324–8. https://doi.org/10.1021/nl1007079.
[31] Hudiono Y, Choi S, Shu S, Koros WJ, Tsapatsis M, Nair S. Porous layered oxide/Nafion® nanocomposite membranes for direct methanol fuel cell applications. *Microporous Mesoporous Mater* 2009;118:427–34. https://doi.org/10.1016/j.micromeso.2008.09.017.
[32] Gahlot S, Kulshrestha V. Dramatic improvement in water retention and proton conductivity in electrically aligned functionalized CNT/SPEEK nanohybrid PEM. *ACS Appl Mater Interfaces* 2015;7:264–72. https://doi.org/10.1021/am506033c.
[33] Hwang S, Lee H, Jeong Y-G, Choi C, Hwang I, Song S, et al. Polymer electrolyte membranes containing functionalized organic/inorganic composite for polymer electrolyte membrane fuel cell applications. *Int J Mol Sci* 2022;23:14252. https://doi.org/10.3390/ijms232214252.
[34] Steffy NJ, Parthiban V, Sahu AK. Uncovering Nafion-multiwalled carbon nanotube hybrid membrane for prospective polymer electrolyte membrane fuel cell under low humidity. *J Memb Sci* 2018;563:65–74. https://doi.org/10.1016/j.memsci.2018.05.051.
[35] He S, Jia H, Lin Y, Qian H, Lin J. Effect of clay modification on the structure and properties of sulfonated poly(ether ether ketone)/clay nanocomposites. *Polym Compos* 2016;37:2632–8. https://doi.org/10.1002/pc.23457.
[36] Du L, Yan X, He G, Wu X, Hu Z, Wang Y. SPEEK proton exchange membranes modified with silica sulfuric acid nanoparticles. *Int J Hydrogen Energy* 2012;37:11853–61. https://doi.org/10.1016/j.ijhydene.2012.05.024.
[37] Ayyaru S, Dharmalingam S. Improved performance of microbial fuel cells using sulfonated polyether ether ketone (SPEEK) TiO_2–SO_3H nanocomposite membrane. *RSC Adv* 2013;3:25243–52. https://doi.org/10.1039/c3ra44212h.
[38] Sasikala S, Meenakshi S, Bhat SD, Sahu AK. Functionalized bentonite clay-sPEEK based composite membranes for direct methanol fuel cells. *Electrochim Acta* 2014;135:232–41. https://doi.org/10.1016/j.electacta.2014.04.180.
[39] Jaafar J, Ismail AF, Matsuura T, Nagai K. Performance of SPEEK based polymer–nanoclay inorganic membrane for DMFC. *J Memb Sci* 2011;382:202–11. https://doi.org/10.1016/j.memsci.2011.08.016.

[40] Shao Z, Xu H, Li M, Hsing I. Hybrid Nafion–inorganic oxides membrane doped with heteropolyacids for high temperature operation of proton exchange membrane fuel cell. *Solid State Ion* 2006;177:779–85. https://doi.org/10.1016/j.ssi.2005.12.035.

[41] Banerjee S, Kar KK. Synergistic effect of aluminium phosphate and tungstophosphoric acid on the physicochemical properties of sulfonated poly ether ether ketone nanocomposite membrane. *J Appl Polym Sci* 2016;133:42952–62. https://doi.org/10.1002/app.42952.

[42] Banerjee S, K. Kar K. Particulate Filled Polymer Electrolyte Membrane for Fuel Cell Applications. *Recent Pat Mater Sci* 2014;7:131–50.

[43] Rangasamy VS, Thayumanasundaram S, De Greef N, Seo JW, Locquet J-P. Preparation and characterization of composite membranes based on sulfonated PEEK and $AlPO_4$ for PEMFCs. *Solid State Ion* 2012;216:83–9. https://doi.org/10.1016/j.ssi.2012.03.017.

[44] Xu C, Wu X, Wang X, Mamlouk M, Scott K. Composite membranes of polybenzimidazole and caesium-salts-of-heteropolyacids for intermediate temperature fuel cells. *J Mater Chem* 2011;21:6014–20. https://doi.org/10.1039/c1jm10093a.

[45] Chien H-C, Tsai L-D, Huang C-P, Kang C, Lin J-N, Chang F-C. Sulfonated graphene oxide/Nafion composite membranes for high-performance direct methanol fuel cells. *Int J Hydrogen Energy* 2013;38:13792–801. https://doi.org/10.1016/j.ijhydene.2013.08.036.

[46] Cao Y-C, Xu C, Wu X, Wang X, Xing L, Scott K. A poly (ethylene oxide)/graphene oxide electrolyte membrane for low temperature polymer fuel cells. *J Power Sources* 2011;196:8377–82. https://doi.org/10.1016/j.jpowsour.2011.06.074.

[47] Shaari N, Kamarudin SK. Graphene in electrocatalyst and proton conductiong membrane in fuel cell applications: An overview. *Renewable Sustainable Energy Rev* 2017;69:862–70. https://doi.org/10.1016/j.rser.2016.07.044.

[48] Wu X, Mu F, Zhao H. Recent progress in the synthesis of graphene/CNT composites and the energy-related applications. *J Mater Sci Technol* 2020;55:16–34. https://doi.org/10.1016/j.jmst.2019.05.063.

[49] Nicotera I, Coppola L, Simari C. Novel Nafion nanocomposite membranes embedded with TiO_2-decorated MWCNTs for high-temperature/low relative humidity fuel cell systems. *Mater Renew Sustain Energy* 2024;13:307–18. https://doi.org/10.1007/s40243-024-00266-7.

[50] Zarrin H, Higgins D, Jun Y, Chen Z, Fowler M. Functionalized graphene oxide nanocomposite membrane for low humidity and high temperature proton exchange membrane fuel cells. *J Phys Chem C* 2011;115:20774–81. https://doi.org/10.1021/jp204610j.

[51] Liang X, Zhang F, Feng W, Zou X, Zhao C, Na H, et al. From metal–organic framework (MOF) to MOF–polymer composite membrane: Enhancement of low-humidity proton conductivity. *Chem Sci* 2013;4:983–92. https://doi.org/10.1039/C2SC21927A.

[52] da Trindade LG, Borba KMN, Zanchet L, Lima DW, Trench AB, Rey F, et al. SPEEK-based proton exchange membranes modified with MOF-encapsulated ionic liquid. *Mater Chem Phys* 2019;236:121792–802. https://doi.org/10.1016/j.matchemphys.2019.121792.

[53] Nur H, Lee Kee G, Hamdan H, Indra Mahlia TM, Efendi J, Cornelis Metselaar HS. Organosulfonic acid functionalized zeolite ZSM-5 as temperature tolerant proton conducting material. *Int J Hydrogen Energy* 2012;37:12513–21. https://doi.org/10.1016/j.ijhydene.2012.06.050.

[54] Yu DM, Yoon YJ, Kim T-H, Lee JY, Hong YT. Sulfonated poly(arylene ether sulfone)/sulfonated zeolite composite membrane for high temperature proton exchange membrane fuel cells. *Solid State Ion* 2013;233:55–61. https://doi.org/10.1016/j.ssi.2012.12.006.

[55] Zeng H, Li Y, Wang C. Using zeolite imidazole frameworks as ionic cross-linkers to improve the performance of composite SPEEK membrane. *Sep Purif Technol* 2020;236:116258–66. https://doi.org/10.1016/j.seppur.2019.116258.

[56] Banerjee S, Kar KK. Superior water retention, ionic conductivity and thermal stability of sulfonated poly ether ether ketone/polypyrrole/aluminum phosphate nanocomposite based polymer electrolyte membrane. *J Environ Chem Eng* 2016;4:299–310. https://doi.org/10.1016/j.jece.2015.11.033.

[57] Banerjee S, Kar KK. Aluminum-substituted phosphotungstic acid/sulfonated poly ether ether ketone nanocomposite membrane with reduced leaching and improved proton conductivity. *High Perform Polym* 2016;28:1043–58. https://doi.org/10.1177/0954008315614984.

[58] Banerjee S, Kar KK. Impact of degree of sulfonation on microstructure, thermal, thermomechanical and physicochemical properties of sulfonated poly ether ether ketone. *Polymer* 2017;109:176–86. https://doi.org/10.1016/j.polymer.2016.12.030.

[59] Hooshyari K, Javanbakht M, Salarizadeh P, Bageri A. Advanced nanocomposite membranes based on sulfonated polyethersulfone: Influence of nanoparticles on PEMFC performance. *J Iran Chem Soc* 2019;16:1617–29. https://doi.org/10.1007/s13738-019-01638-x.

[60] Anilkumar GM, Nakazawa S, Okubo T, Yamaguchi T. Proton conducting phosphated zirconia–sulfonated polyether sulfone nanohybrid electrolyte for low humidity, wide-temperature PEMFC operation. *Electrochem Commun* 2006;8:133–6. https://doi.org/10.1016/j.elecom.2005.10.025.

[61] Zhou W, Xiao J, Chen Y, Zeng R, Xiao S, Nie H, et al. Sulfonated carbon nanotubes/sulfonated poly(ether sulfone ether ketone ketone) composites for polymer electrolyte membranes. *Polym Adv Technol* 2011;22:1747–52. https://doi.org/10.1002/pat.1666.

[62] Gosalawit R, Figoli A, Chirachanchai S. Physicochemical and electrochemical characterizations of organic montmorillonite (OMMT)/sulfonated poly(ether ether ketone) (SPEEK) composite membranes. *Asia-Pacific J Chem Eng* 2010;5:60–5. https://doi.org/10.1002/apj.389.

[63] Hasani-Sadrabadi MM, Dashtimoghadam E, Sarikhani K, Majedi FS, Khanbabaei G. Electrochemical investigation of sulfonated poly(ether ether ketone)/clay nanocomposite membranes for moderate temperature fuel cell applications. *J Power Sources* 2010;195:2450–6. https://doi.org/10.1016/j.jpowsour.2009.11.090.

[64] Ali MM, Rizvi SJA, Azam A. Fabrication of proton exchange membranes and effect of sulfonated SiO_2 (S-SiO_2) in sulfonated polyether ether ketone (SPEEK) for fuel cells applications. *IOP Conf Ser Mater Sci Eng* 2019;577:012039–47. https://doi.org/10.1088/1757-899X/577/1/012039.

[65] Gohel JV, Mishra PS, Murthy ZVP. TiO_2 nanoparticles prepared by mechanical reduction technique for superior DMFC nanocomposite PVA membranes. *Sep Sci Technol* 2019;54:233–46. https://doi.org/10.1080/01496395.2018.1501064.

[66] Xu T, Hou W, Shen X, Wu H, Li X, Wang J, et al. Sulfonated titania submicrospheres-doped sulfonated poly(ether ether ketone) hybrid membranes with enhanced proton conductivity and reduced methanol permeability. *J Power Sources* 2011;196:4934–42. https://doi.org/10.1016/j.jpowsour.2011.02.017.

[67] Khalid F, Roy AS, Parveen A, Castro-Muñoz R. Fabrication of the cross-linked PVA/TiO_2/C nanocomposite membrane for alkaline direct methanol fuel cells. *Mater Sci Eng: B* 2024;299:116929–36. https://doi.org/10.1016/j.mseb.2023.116929.

[68] Di Y, Yang W, Li X, Zhao Z, Wang M, Dai J. Preparation and characterization of continuous carbon nanofiber-supported SPEEK composite membranes for fuel cell application. *RSC Adv* 2014;4:52001–7. https://doi.org/10.1039/C4RA06976E.

[69] Zhnag Shu. Mechanical and physical properties of electrospun nanofibers. North Carolina State University, 2009.

[70] Shirin -KK S, Kandasubramanian B. Sulfonated PTFE and its composites for high-temperature fuel cell applications: A review. *Thermal Adv* 2025;3:100043–54. https://doi.org/10.1016/j.thradv.2025.100043.

[71] Barbosa JC, Pinto RS, Correia DM, Fidalgo-Marijuan A, Silva MM, Gonçalves R, et al. Effect of fluorinated polymer matrix type in the performance of solid polymer electrolytes based on ionic liquids for solid-state lithium-ion batteries. *Chem Eng J* 2023;478:147388–14797. https://doi.org/10.1016/j.cej.2023.147388.

[72] Li M, Scott K. A fluorinated polymer/inorganic composite electrolyte membrane for intermediate temperature fuel cells. *Int J Energy Res* 2012;36:1000–4. https://doi.org/10.1002/er.1876.

[73] Zhang X, Trieu D, Zheng D, Ji W, Qu H, Ding T, et al. Nafion/PTFE composite membranes for a high temperature PEM fuel cell application. *Ind Eng Chem Res* 2021;60:11086–94. https://doi.org/10.1021/acs.iecr.1c01447.

[74] Yoon KR. Cross-aligned PTFE nanofiber reinforced composite membrane for PEM fuel cells. *ECS Meeti Abstr* 2023;MA2023-02:1901–1901. https://doi.org/10.1149/MA2023-02391901mtgabs.

[75] Schuster MFH, Meyer WH. Anhydrous proton-conducting polymers. *Annu Rev Mater Res* 2003;33:233–61. https://doi.org/10.1146/annurev.matsci.33.022702.155349.

[76] Chen J, Cao J, Zhang R, Zhou J, Wang S, Liu X, et al. Modifications on promoting the proton conductivity of polybenzimidazole-based polymer electrolyte membranes in fuel cells. *Membranes* 2021;11:826–48. https://doi.org/10.3390/membranes11110826.

[77] Díaz M, Ortiz A, Ortiz I. Progress in the use of ionic liquids as electrolyte membranes in fuel cells. *J Memb Sci* 2014;469:379–96. https://doi.org/10.1016/j.memsci.2014.06.033.

[78] Aili D, Henkensmeier D, Martin S, Singh B, Hu Y, Jensen JO, et al. Polybenzimidazole-based high-temperature polymer electrolyte membrane fuel cells: New insights and recent progress. *Electrochem Energy Rev* 2020;3:793–845. https://doi.org/10.1007/s41918-020-00080-5.

[79] Hooshyari K, Rezania H, Vatanpour V, Salarizadeh P, Askari MB, Beydaghi H, et al. High temperature membranes based on PBI/sulfonated polyimide and doped-perovskite nanoparticles for PEM fuel cells. *J Memb Sci* 2020;612:118436–51. https://doi.org/10.1016/j.memsci.2020.118436.

[80] Gorre A, Das A, Jana T. Mixed matrix composite PEM with super proton conductivity developed from ionic liquid modified silica nanoparticle and polybenzimidazole. *J Macromol Sci, Part A* 2023;60:38–50. https://doi.org/10.1080/10601325.2022.2154677.f

6 Recent developments in non-Nafion PEMFCs

6.1 HYDROCARBON-BASED PEM

PEMFCs can be broadly classified into two classes such as cationic and anionic exchange membranes. For PEMFCs, the most extensively used commercial cation exchange membrane is based on perfluorosulfonic acid (PFSA) membranes, commonly called the Nafion® membranes. As discussed in our earlier chapter, Nafion® has a unique morphological structure where hydrophilic and hydrophobic domains are present within the main polymeric backbone and are the main responsible unit for high proton conduction in this membrane. However, due to high cost and other drawbacks, scientists are constantly working for the development of suitable alternative membranes. Hydrocarbon based membranes are the most common in this respect due to several advantages over the PFSA membranes, however the proton conductivity still remains lower than commercial Nafion® that requires further research work. Hydrocarbon polymers are easy to synthesize and are of low cost with tuneable proton conductivity that make them material of interest [1]. However, low thermal stability and chemical resistance make them unsuitable as PEM during the early research period. Modified hydrocarbon based PEMs are low cost and have the potential to be utilized for small scale portable application where lifetime requirement is much less as compared to the space vehicles and heavy end applications. Also, researchers are working tirelessly to develop catalyst and design MEA assemblies that can generate maximum efficiency at low cost.

Hydrocarbon polymers are of various advantages such as:

a. They are cheaper as compared to commercial PFSA membranes.
b. Easy to synthesize with provision of tailoring the design of polymeric backbone as per requirement.
c. The functionalized hydrocarbon polymers contain polar groups that make them hydrophilic and thereby assisting the proton conduction.
d. The dimensional stability and hydrophilicity of these membranes can be easily controlled by functionalization and crosslinking or grafting that opens up suitability of these materials for various applications.
e. Hydrocarbon polymers can be recycled by conventional processes that are definitely valuable in term of environmental protection and sustainability.

However, the main concern remains the balance between the hydrophilicity, dimensional stability, and proton conductivity since these membranes suffer from the problem of excessive swelling at higher degree of functionalization. The success of

DOI: 10.1201/9781003733744-6

these membranes is very much dependent on the microstructure and hence extensive research is going on to tailor the microstructure as per requirement. The microstructure of these membranes can be varied by blending, crosslinking, functionalization, grafting, etc. [2]. In this chapter, we will discuss various types of membranes that are of recent research interest to give a broad overview on the non-Nafion® type of PEM materials for low-temperature fuel cells.

6.2 POLYSULFONE PEM

Polysulfones serve as attractive PEM materials for fuel cells due to the excellent thermal and oxidative stability, mechanical properties, great hydrolytic stability, ease of processing, and high glass transition temperature. However, to be an efficient PEM material, it requires a high degree of sulfonation which leads to increased fuel crossover, low dimensional stability in water or methanol, and finally inferior mechanical properties in the membranes. Sulfonation of polysulfones is advantageous due to its high mechanical, chemical, thermal and oxidative stability, good proton conductivity, methanol resistivity, wide availability, and low cost. However, they suffer from two major drawbacks, first, excessive swelling due to the presence of hydrophilic sulfonic acid groups and, second, decrease in conductivity due to the degradation of sulfonic acid side chains specifically at high temperatures. With the aim to fabricate an alternative to Nafion, researchers have developed polysulfone-based membranes having different degrees of functionalization as well as filled with various inorganic fillers [3]. Inorganic fillers, such as silica, heteropolyacids, clays, metal oxides (titania and zirconia), and carbon nanotubes, are the most common fillers incorporated to improve the functional properties of the PEMs [4]. Ozden et al. have investigated sulfonated polysulfone composite membranes filled with inorganic filler zirconium hydrogen phosphate (ZrP) [5]. The sulfonation degree of the polymer is varied as 20%, 35%, and 42%. A fixed amount of ZrP (2.5 wt.%) has been incorporated to investigate its effect on the membrane properties, particularly pertaining to DMFC application and compared with the commercial Nafion 115 membranes. In another work, Kim et al. have designed a comb-shaped polysulfone copolymer by the click reaction of polysulfones containing azidomethyl side groups and sulfonated polytriazole bearing one ethynyl chain-end groups [6]. The ion exchange capacity of the membranes is controlled by varying the chain length of sulfonated polytriazoles. The tough and flexible membranes have exhibited significant proton conduction ability and physicochemical properties with excellent fuel cell performance due to the distinct hydrophilic-hydrophobic phase separation. An acid-base interaction between the sulfonic acid and triazole groups of the copolymer promotes outstanding mechanical strength as compared to the linear polymer membrane. The cell performance of these membranes is reported to be superior as compared to the commercial Nafion 212 membrane at different operating conditions.

With the aim to fabricate PEM of high conductivity and improved methanol permeability, composite membranes are fabricated with various inorganic fillers such as clay, zeolite, and inorganic oxides. ZSM-5 is a hydrated aluminosilicate used to improve the proton conductivity, mechanical properties, and water retention ability of PEM. The hydrated aluminosilicate is composed of alumina and silica consisting of intra-crystalline channels and cavities enabling adsorption and desorption of organic molecules. In recent work, Krathumkhet et al. have developed polysulfone composite membranes using ZSM-5 for DMFC applications [7]. The composite membranes are characterized by improved proton conductivity, water uptake, ion exchange capability, and methanol permeability. The degree of sulfonation of the polysulfone is about 70%. A ternary nanocomposite membrane made of sulfonated polysulfones, silica nanoparticles, and metal-organic framework (MOF) has also been investigated [8]. The nanocomposite membranes have exhibited a power density of 40.80 mW cm^{-2}. The properties of these membranes are reported to be much better than the pristine polysulfone blend membranes. Kim et al. have studied sulfonated poly(arylene ether sulfone) (sPES) composite membranes filled with mesoporous ceria-silica inorganic filler and fabricated by solution-casting approach [9]. The addition of the composite filler improves the oxidative stability of these membranes however, a decrease in proton conductivity has been evidenced which has to be resolved further by suitable modification of the inorganic filler or by modifying the structure of the base polymer matrix.

6.3 POLYPHOSPHAZENE PEM

Polyphosphazenes are the alternative class of PEM consisting repeating unit with alternating nitrogen and phosphorous atoms. The properties of polyphosphazenes are largely dependent on the chemical structure of side chains. These polymers have an interesting combination of properties such as high thermal and chemical stability since both the nitrogen and phosphorous atoms are present in their highest oxidation states contributing to a high degree of thermo-oxidative stability. Again, the polar nature of bonding provides resistance against the attack of free radicals. Finally, the absence of long conjugation makes this class of polymers highly flexible. Polyphosphazenes with aryloxy substitution possess high chemical, thermal and mechanical stability and hydrophobic characteristic, hence provide an excellent combination of properties. Again, these polymers can offer improvement in mechanical and barrier performance due to the ease of crosslinking generally via methyl groups in the presence of gamma radiation.

Polyphosphazene-based polymer electrolyte membrane is one of the attractive classes of membrane material suitable for both hydrogen and methanol-based fuel cell devices. The cation-exchange membranes are known for their proven thermal and chemical stability due to the ease of structural modification through side-chain attachment and crosslinking in the polymeric backbone [10, 11]. Polyphosphazene has also been grafted with SPEEK to prepare PEM. Zhao et al. have developed poly(sulfonic acid)cyclotriphosphazene derivative grafted with SPEEK having a sulfonation degree of 65% to improve the proton conductivity of the fuel cells [12]. The grafted membranes are characterized by improved proton conductivity in the order of 10^{-2} S cm^{-1} at a grafting degree of 15%. The membranes have exhibited methanol permeability as low as 2.72×10^{-7} cm^{-2} s^{-1} which is quite useful for DMFC devices. In another work, Dong et al. have studied PEM based on sulfonated polyphosphazene containing pendant perfluorosulfonic acid (PMFP-g-PS) groups and SPEEK for DMFC applications [13]. The crosslinked membranes have been prepared in a stepwise manner.

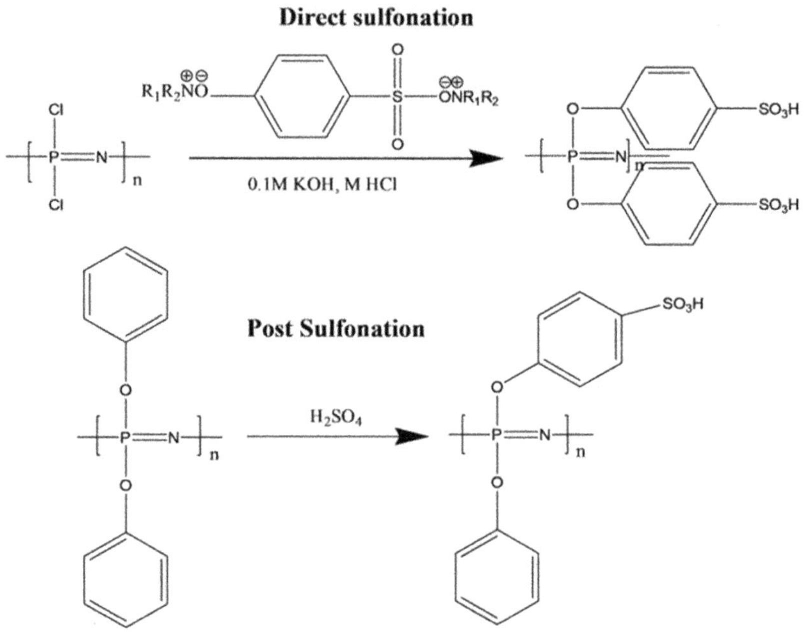

FIGURE 6.1 Reaction mechanism of direct and post-sulfonation of polyphosphazene

The synthesized membranes have exhibited much lower methanol permeation along with considerably high selectivity as compared to commercial Nafion 117 membranes. Membranes also have exhibited a good combination of properties such as high dimensional stability, low water uptake, outstanding proton conductivity, and low methanol permeation, good thermal and oxidative stability which prove their suitability as PEM materials for the low temperature fuel cell devices. The microstructural and proton conduction properties of sulfonated poly[bis(3-methylphenoxy)phosphazene] are studied by small and wide-angle X-ray diffraction techniques [14]. The structural investigation revealed that the membrane consists of an ordered phase and an amorphous phase. The 2D structure of the membranes changes dramatically with the extent of water swelling and ion exchange capacity. The cluster size of hydrated films also shows a similar trend. A polyphosphazene-based polymer can only be used as PEM material when it contains any kind of protogenic functionalization enabling proton conduction through the membrane. The most common functionalization is the sulfonation like other synthetic polymers either through direct sulfonation or by post-sulfonation of the polymer (Figure 6.1). The post-sulfonation process is easier however, direct synthesis of functionalized polymer would always provide better control over the polymer properties due to the minimization of unwanted side reaction and degradation of molecular weight.

Acidity in polyphosphazenes is introduced by functionalization with sulfonic and phosphonic acid groups. The membrane properties such as proton conductivity, water content, and methanol permeation, are greatly affected by the nature of functionalization present in the polymer [15]. Sulfonation of polyphosphazenes imparts high acidity and thermal stability. A considerable degree of sulfonation say as high as 30% of the side groups induce water solubility, however, the extent of swelling of the polymers can be controlled by the judicial choice of crosslinking. Principally for phosphonated polyphosphazenes the extent of swelling or uptake of water are quite less. For each phosphonic acid unit, there are two acidic protons. However, due to weaker acidic character of the phosphonic acid groups with respect to the sulfonic acid groups they are less hydrophilic as compared to the sulfonic acid groups. A group of studies have shown that the methanol permeation of phosphonated polyphosphazenes is about nine times lower as compared to commercial Nafion at 120°C temperature which make them suitable for DMFC application [16]. Sulfonimide group of polymers are well known for their high acidity [17]. Sulfonimide substituted polyphosphazenes are suitable as PEM material. Polyphosphazene copolymer bearing sulfonimide pendant groups and 4-methylphenoxy groups have been studied for their fuel cell properties [18]. The polymer containing 17% sulfonimide and 83% *p*-methylphenoxy groups has achieved IEC of 0.99 meqv g^{-1} which is close to 95% of the theoretical maximum. The membranes also exhibited proton conductivity as high as 4.9×10^{-2} S cm^{-1}, however, the water uptake property of these membranes was quite high (about 119 wt.%). The water uptake of these membranes is reduced by two methods. In a typical process, they are blended with polyvinylidene fluoride (PVDF) in a ratio of polyphosphazene: PVDF= 3:1 has shown a reduced water uptake of 41% with proton conductivity as high as 6.0×10^{-2} S cm^{-1}. In another strategy, the membranes are treated with gamma radiation in doses 20 and 40 Mrad to impart crosslinking. The water uptake of these membranes after treatment reduces to 73 and 4 wt.%, respectively.

6.4 SULFONATED POLYIMIDE (SPI) PEM

A new class of PEM is developed based on polyimides [19, 20]. Polyimides exhibit attractive properties such as reduced sensitivity to environmental condition, lamellar or disk-type microstructure in membrane as well as in the case this polymer water content per ionic group is independent of the equivalent group of the polymer. However, poor solubility of polyimides limits their applicability. Polyimides are characterized by fairly high thermal stability (stable up to 300°C) which facilitates the formation of compatible nanocomposite membranes with ceramic materials. Excellent thermal, mechanical, and chemical stability of polyimide along with good insulating properties make them, a material of choice for modern applications. Sulfonation of polyimide imparts good proton conductivity and hydrophilicity that are necessary prerequisite for polymer electrolyte membranes. Sulfonated polyimides are a good alternative to perfluorosulfonic acid (PFSA) membranes for fuel cells due to their high proton conduction ability in the hydrated state along with good mechanical properties and dimensional stability. Polyimides can be sulfonated by two ways, side chain and main chain sulfonation. Sulfonated polyimides also undergo swelling and require hydration for proton conduction similar to Nafion. Moreover, properties like water uptake, electro-osmotic drag coefficient, and swelling ratio, do not change over a long range of temperature like Nafion. The proton conductivity of sulfonated polyimides have been reported to be in the range of

2×10^{-3} to 4×10^{-2} S cm^{-1} for the first generation, which is lower than Nafion [21, 22]. For next-generation, SPI membranes conductivity reaches up to 1.67 S cm^{-1} at 120°C and 1.201 S cm^{-1} at 80°C for graphene-based nanocomposite membranes [23, 24]. Sulfonated polyimide membranes are also characterized by low methanol permeability as compared to Nafion specifically at higher temperatures which makes them a potential candidate for DMFC applications. The main drawback of the sulfonated polyimide-based membrane is the lack of stability in the hydrated state, which influence various aspects such as hydrolytic and oxidative stability, stress, and swelling. The imide linkage present in the chemical structure of polyimide is prone to hydrolysis. This leads to main chain scission in polyimides with a reduction in average chain length of polymer molecules [25]. The reduction in chain length leads to mechanical degradation of the polymer backbone which is reflected by cracking and increased gas permeability in the membrane limiting its application in fuel cell devices.

The hydrolysis of imide occurs due to the nucleophilic interaction between the water molecules and carbon atoms of the carbonyl groups (Figure 6.2). Therefore, the molecules with more electron density at the carbonyl carbon will impart more stability against hydrolysis of imide. Based on this theoretical consideration, hydrolysis of polyimide has been prevented by various routes. The electron density at carbonyl carbon has been improved by converting the five-membered heterocycles to a six-membered one by conversion of phthalic polyimides to naphthalenic polyimides [26, 27]. Another way to improve the hydrolytic stability is to introduce bulky aliphatic groups either in the main chain or as pendant functional groups in the side chain [28, 29]. The electron density at the carbonyl carbon has also been enhanced by introducing bridging sulfur electron donor in the main polymer chain. The sulfur

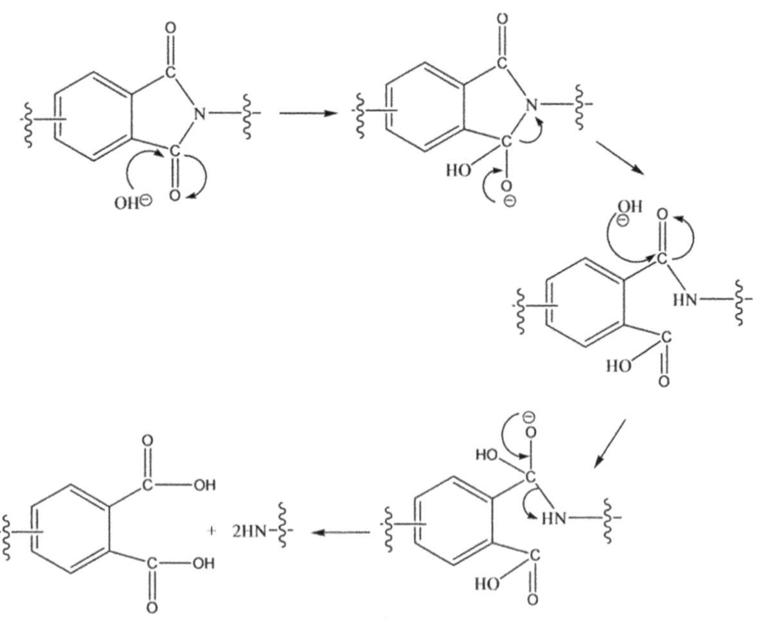

FIGURE 6.2 Hydrolytic degradation of polyimide

lowers the electrophilicity of carbonyl carbon and hence imparts resistance against hydrolysis [30, 31]. The other possible route to improve the susceptibility to hydrolysis is to introduce electron-donating phenoxy groups [32].

The proton conductivity of sulfonated polyimide increases with the increase in water content [33]. However, at low water content, the proton conductivity of sulfonated polyimide is lower as compared to the Nafion. However, even at low water content, Nafion retains substantial small water channels, which facilitate better proton conduction as compared to sulfonated polyimide. The proton conductivity of sulfonated polyimide is found to be improved with the increase in ion exchange capacity, i.e., the density of sulfonic acid groups. Further, the absorption of water enhances with the increase in ion exchange capacity. The performance of sulfonated polyimide membranes is determined by testing under practical fuel cell operating conditions however, the final efficiency is fairly dependent on the stability of the membrane as well as stability of the catalyst/membrane interface.

6.5 POLYSTYRENE-BASED PEM

Polystyrene sulfonic acids (PSSA) are the first commercial polymer electrolyte membranes. Although short lifetime and severe degradation of the membranes and introduction of commercial Nafion membrane hinder their practical application however, good proton conductivity at reduced relative humidity and higher temperature still make them significant for their application in PEM. PSSA membranes also possess better methanol selectivity as compared to the Nafion [34]. However, the major drawback of PSSA membranes is the rapid loss of SSA fragment under operating conditions resulting in a sudden drop in membrane performance with the time. PSSA polymer chains consist of the weak point at tertiary benzylic hydrogen as well as aromatic ring protons. Hence, peroxide can attack at those hydrogens leading to severe degradation [35]. The stability of PSSA membranes is found to be improved by eliminating tertiary hydrogens via substitution with methyl and fluorine groups [36]. Stable sulfonated poly(α,β,β-trifluorostyrene) membranes were developed by Assink and Arnold to improve the energy efficiency under oxidative environment. Energy efficiencies of the substituted membranes were observed to be 82.6% and 83% under 700 test cycles with no chemical degradation indicating high oxidative stability of the membranes under harsh conditions [36]. The water content of PSSA membranes is much higher as compared to Nafion, however conductivity is quite low. This may be attributed to the presence of small and less interconnected water channels in the PSSA membranes. Again, low proton conductivity value of these membranes is also expected due to the presence of lesser number of sulfonic acid groups. However, low electro-osmotic pressure drag is advantageous for these membranes. Various SSA copolymers with different chain lengths have been prepared and studied for fuel cell properties as the properties of PSSA membranes are dependent on the nature of the polymer backbone. Triblock copolymers, block copolymers, sulfonated styrene–ethylene copolymers, poly-[norbornenyl ethyl styrene-*s*-styrene]-poly (*n*-propyl-*p*-styrene sulfonate) block polymers, etc., are prepared and investigated for fuel cell properties [37, 38]. These membranes are characterized by high proton conductivity and methanol selectivity as compared to the Nafion; however, the stability of these membranes in the oxidative environment is the main impediment limiting the use of these membranes in practical fuel cell applications.

The block copolymer membrane made of SSA suffers from serious swelling problem, which limits the achievement of high IEC and hence leading to the low proton conductivity value in this membrane. The stability of PSSA membranes against water has been accomplished by grafting of SSA moiety onto the backbone of a stable polymer. Grafting in this way prevents excessive swelling and hence high IEC and conductivity can be achieved with moderate mechanical stability. SSA and modified SSA moieties are grafted onto the backbone of stable polymers such as PTFE, PVDF, PVDF-co-HFP, and PVF, by radiation grafting technique [39–42]. Among these, the PVDF-grafted-SSA membrane exhibits good proton conductivity of 0.13 S cm^{-1}. However, very high water absorption, about 59 molecules of water per sulfonic acid group, limits its application as the PEM [43]. Grafting with α, β, β-trifluorostyrene and α-methylstyrene instead of SSA fragment exhibits enhanced stability with no degradation up to 1053 hours of testing at 60°C in a fuel cell test station [44]. However, low grafting rate of these membranes is a major problem. To overcome this problem, SSA has been grafted with various monomers such as methacrylonitrile, methyl methacrylate, methacrylic acid, and acrylonitrile.

SSA fragment has also been blended with stable polymeric backbones. Within a blend of PVDF-co-HFP/Nafion, sulfonated styrene in salt form has been impregnated and subsequently in-situ polymerized [45]. The cross-linked semi-interpenetrating polymeric membrane exhibits fuel cell properties comparable to the standard Nafion. An optimum amount of sulfonated styrene (20 wt.%) has exhibited a maximum current density of 120 mA cm^{-2} at 0.2 V and a power density of 24 mW cm^{-2} at 60°C. The membrane has achieved desired physicochemical properties such as water uptake, proton conductivity, membrane selectivity, and current density, comparable to commercial Nafion membrane indicating its potential as PEM for DMFC devices. In a similar system, Piyush et al. have investigated the effect of temperature on the electrochemical cell performance of PEM [46]. The sulfonated styrene is beneficial in maintaining the hydration state of the membrane as proper retention of moisture is required to facilitate the electrochemical performances. These membranes are reported to exhibit decent thermal stability and low interfacial resistance allowing them to work at elevated temperature.

An inexpensive phase-separated PEM with tunable macromolecular designs has been developed by Sodeye et al. [47]. These membranes have been prepared from block copolymers of fluorinated polyisoprene-block-sulfonated polystyrene at various degrees of sulfonation. With an increase in the degree of sulfonation, the transport properties of membranes are found to be improved. Further, the study reveals that the membranes in acid form are capable of absorbing more water than the one in cesium form. These membranes have also exhibited more than an order of less methanol permeability as compared to Nafion 112 establishing their suitability for DMFC application. Sulfonated polystyrene grafted polypropylene composite membranes have also studied for fuel cells [48]. The composite membranes have been prepared by grafting polystyrene on to microporous polypropylene membranes via plasma-treated polymerization followed by sulfonation. The composite membranes have exhibited proton conductivity as high as 0.019 S cm^{-1}. In summary, to date, the performance of PSSA membranes remain quite low as compared to the Nafion membrane in terms of conductivity (18 mW cm^{-2} for PSSA *vs.* 31 mW cm^{-2} for Nafion) despite their low methanol permeability [49].

6.6 ELECTEROSPUN FIBER-BASED PEM

Membranes obtained by electrospinning are reported to be highly proton conductive as compared to the solution casted one. A new path of research has been opened up for Nafion based PEM via the development of electrospun Nafion nanofibers-based membranes.

Inorganic-organic nanocomposite polymer electrolytes have been extensively studied in the literature. Carbon nanofibers (CNF) with tunable nanostructures can be prepared by electrospinning. Di et al. have studied continuous CNF (CCNF) mats as a filler to prepare composite PEM [50]. The results suggest that the incorporation of CCNF is effective in improving various fuel cell related properties such as water uptake, mechanical properties, and proton conductivity. However, a decrease in methanol selectivity has been evidenced in this membrane. Liu et al. have prepared sulfonated CNF (SCNF)/SPEEK nanocomposite PEM for fuel cells [51]. CNF mat has been prepared by electrospinning followed by sulfonation to obtain SCNF. Excellent bonding between the SCNF and SPEEK is obtained by hydrogen bonding among the sulfonic acid groups of SPEEK and SCNF containing water molecules. Comparative analysis of the result suggests that, neat SPEEK membrane without SCNF possesses much lower proton conductivity and methanol permeability. The composite membranes exhibit good water uptake, electrochemical properties, and mechanical strength. However, a very high amount of SCNF in the polymer matrix may lead to aggregation of the filler and thereby reducing the proton conductivity of the membranes. The best combination of properties is reported for the composite membranes with 1 wt.% SCNF showing proton conductivity comparable to Nafion 117 (0.128 S cm^{-1}) with substantially lower methanol permeability of 5.12×10^{-7} cm^2 s^{-1}.

A good PEM should exhibit good proton conductivity along with low methanol permeability. To achieve these properties, various electrospinning parameters such as the concentration of solution, voltage, humidity, and distance, need to be controlled. However, one of the common drawbacks of electrospun fibers is their low mechanical strength. They tend to break down easily due to their small diameter [52]. Hence a reinforcing material capable of improving the mechanical strength of the nanofibers is incorporated to withstand the effect of elevated temperature and pressure during prolonged operation.

6.7 CROSSLINKED PEM

Crosslinking is an effective tool in controlling high swelling of polymers having high degree of sulfonation (DS). High DS is essential to achieve high proton conductivity; however, a very high content of water-loving groups leads to excessive swelling and reduced mechanical property. A sensible introduction of crosslinks into the polymer matrix may lead to better control over the swelling of polymers with high IEC. However, crosslinking in an uncontrolled manner approaching to the gel point leads

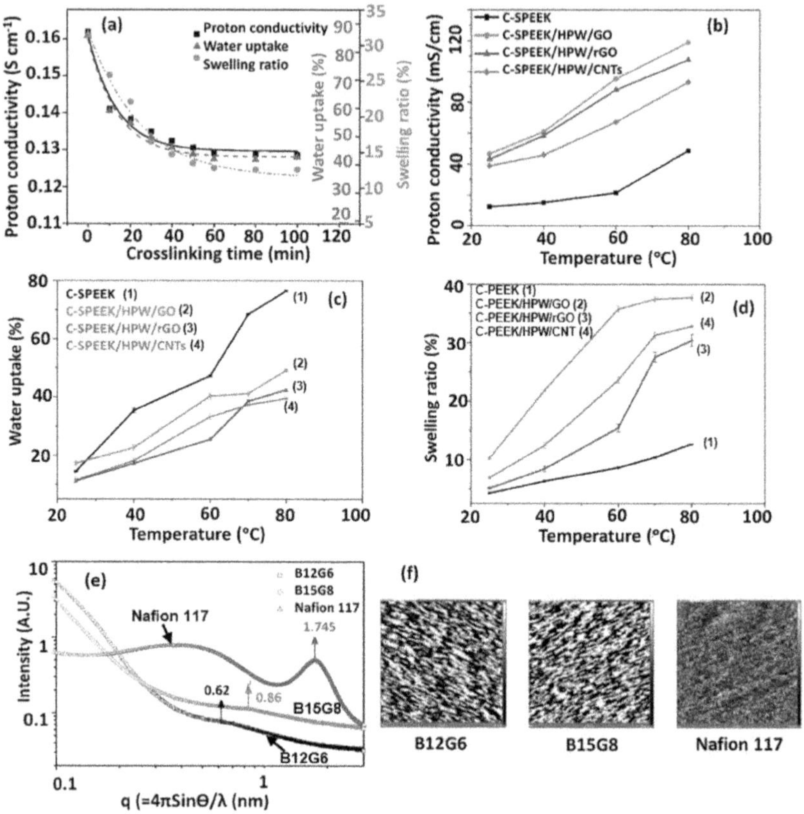

FIGURE 6.3 (a) Effect of cross-linking time in minute on proton conductivity, water uptake, and swelling ratio of membranes [53], variation of (b) proton conductivity, (c) water uptake, and (d) swelling ratio with temperature for self-crosslinked sulfonated Poly(etheretherketone) composite membranes [54]; (e) SAXS spectra and (f) AFM images of Nafion 117 and cross-linked PVA membranes with different crosslinking agents [55]. (Reprinted with permission)

to adverse effects with a huge reduction in proton conductivity. Figure 6.3 shows the effect of crosslinking on the proton conductivity, water uptake, and swelling ratio of the PEM. Crosslinking in polymers can be introduced by three main routes, namely ionic crosslinking, UV-assisted crosslinking, and thermally induced crosslinking.

6.7.1 Ionic Crosslinking

Ionic crosslinking is an effective way to improve the mechanical properties of polymer electrolytes [56, 57]. In this case, crosslinking is obtained by a chemical reaction between the sulfonated polymer bearing the -SO_3H groups and the basic polymer, such as PBI, e.g., SPEEK/PBI system. After casting and acidification of the membranes, formation of an acid-base complex occurs due to a chemical reaction between the -SO_3H and N-H groups. The crosslinking can be controlled by controlling the amount and basicity of the basic polymer. Below the percolation threshold of the acidic polymer, say in the case of sulfonated polymer, most of the acidic groups are ionically trapped in PBI and the proton conductivity remains quite low at this stage. Once the sulfonation degree reaches above the threshold value, the water-soluble nature of sulfonated polymer is converted into insoluble one due to the crosslinking and under this situation it started to exhibit certain conductivity value. Instead of morphological stability, the water swelling in this system remains much higher as compared to those, where crosslinking is based on covalent bonds. The strength of physical crosslinking is not sufficient enough to prevent the excessive swelling over 80°C for such systems. However, the introduction of amphiphilic copolymers containing sulfonic acid groups in addition to PBI as the basic polymer can compensate the loss in proton conductivity with maintained fuel cell performance [58].

6.7.2 UV-assisted crosslinking

Sulfonated polyphosphazenes having low glass transition temperature enable photocrosslinking in the membrane in the presence of UV light [59]. The sulfonated polyphosphazene solution must contain a photo-initiator such as benzophenone which can initiate photocrosslinking in the cast membrane. The study reported substantial improvement in dimensional stability in the crosslinked membrane at the expense of a significant loss in proton conductivity. The crosslinked sulfonated polyphosphazene membranes are considered to be more suitable for DMFC applications as compared to the PEMFC devices due to the low methanol diffusivity. Guo et al. have developed a proton exchange membrane based on photocrosslinked polyphosphazene for PEMFC and DMFC applications [11]. Photocrosslinking has been initiated by the dissolution of benzophenone photoinitiator in the casting solution followed by drying and crosslinking of the resulted film through exposure under the UV light. The membranes have exhibited promising properties and much less swelling than the control Nafion 117 membrane both in water and methanol medium. The membranes have also displayed good mechanical integrity up to 173°C and 800 kPa pressure and excellent chemically stability in hot hydrogen peroxide/ferrous ion solution.

6.7.3 Thermally induced crosslinking

Sulfonated copolymers have been cross-linked by thermal treatment over a wide temperature range of 90–360°C depending on the choice of crosslinking agent and cross-linkable groups. These cross-linked copolymers can be classified into two groups, first, the polymer and cross linker system and second, the self-cross-linkable polymeric system. A very common and conventional route of thermal crosslinking is crosslinking between the sulfonic acid groups of sulfonated copolymers and hydroxyl groups of polyatomic alcohols, such as glycerol and ethylene glycol, at a crosslinking temperature range of 120–150°C [60]. The major drawback of this route is a severe reduction in proton conductivity even for polymers with a high degree of sulfonation. Therefore, the inclusion of crosslinkers which are capable of reacting with functional groups other than sulfonic acid groups can be a good choice. Representative cross-linked copolymers are obtained from BPSH (biphenyl disulfonated polyaryl ether sulfone) with phenoxide terminal groups. The phenoxide terminal ends in the BPSH oligomers are reacted with the crosslinker and converted in a cured membranes. The acidification of the crosslinked membranes is done afterward to achieve desired proton conduction property. Glycidyl methacrylate is the common crosslinker used for this kind of system. Tetraglycidyl bis(*p*-aminophenyl) methane is another important crosslinker with better hydrolytic and thermal stability than that of glycidyl methacrylate. A small amount of NMP is used to lower the glass transition temperature of oligomer with high sulfonation degree to ease the process of crosslinking. The crosslinking followed by acidification of BPSH-50 minimizes the water uptake from 232% to 84%. This reduction in water uptake may be useful for the enhancement in proton conductivity by preventing dilution of proton concentration per unit volume of the polymer due to the excessive swelling.

The crosslinking of sulfonated copolymers generally lowers the proton conductivity due to the restriction imparted in chain mobility and blocking of water channels. A good idea to compensate this loss can be the use of a crosslinker with sulfonic acid groups [61]. Polyvinyl alcohol-sulfosuccinic acid (PVA-SSA) system is one of the examples of this kind. In this case, SSA plays the role of a sulfonating agent as well as the crosslinker [62]. After casting of the membrane, hydroxyl groups of PVA react with the carboxylic groups of SSA at a temperature range of 120–130°C via a polycondensation reaction to produce a crosslinked structure. The proton conductivity of the system may be tuned depending on the curing temperature and amount of curing agent, i.e., SSA. Introduction of a crosslinker with sulfonic acid groups such as *N*, *N*-bis(2-hydroxyethyl)-2-aminoethanesulfonic acid (BES) is effective in reducing water uptake while improving the IEC [61]. The additional sulfonic acid groups from the crosslinker side are helpful in improving the ionic density of the polymeric system. The BES crosslinked sulfonated polyimide (SPI) exhibits excellent proton conductivity of 0.1 S cm^{-1} at 90°C at 90% RH with an added advantage of low methanol permeation (1.6×10^{-7} cm^2 s^{-1}). This unusual result on methanol permeation, in respect to the general trend, might be due to the fact that BES reduces the free space capable of absorbing free water molecules in the polymer membrane thereby creating a dense structure. This dense morphology acts as a barrier for methanol permeation as a result of restriction in chain mobility imparted by the sulfonic acid side chains of the BES unit. The reduction in free water gradually lowers the methanol permeation.

Self-cross-linkable polymers are made of homopolymers with thermally cross-linkable side groups such as sulfonic acid and ethynyl groups. At a particular curing temperature range, the cross-linkable functional groups form inter or intramolecular covalent bonds with the reactive sites present in the polymer backbone. In recent studies, self-cross-linkable polymer systems are reported in the literature [53, 63]. SPEEK with sulfonated polyphenylsulfone (SPPSU) is an example of a self-cross-linkable polymeric membrane, where a portion of the sulfonic acid group from the crosslinker site form sulfone bridges during heat treatment to cure the polymer matrix [64]. The crosslinked polymers are characterized by enhanced thermal, mechanical, and dimensional stability. However, during the process of crosslinking, the loss of proton conductivity is unavoidable. The possible way to minimize the reduction in proton conduction can be the use of a highly sulfonated polymer matrix or the use of a crosslinking agent bearing pendant sulfonated groups.

6.8 POLYMER BLENDS-BASED PEM

The high cost and reduction in performance at elevated temperatures are the major drawbacks of Nafion membranes that demand the development of alternative membranes. The development of anhydrous polymer electrolyte membranes that can maintain their performance at elevated temperatures and low humidity is of primary concern. In this context, different strategies have been adopted to achieve efficient proton conduction at elevated temperatures and during prolonged operation. One of the common approaches is the preparation of polymer acid-base complexes based on PBI and H_3PO_4 [65, 66]. In polymer complexes, the structural diffusion is prominent while the diffusion of protons is mainly controlled by the transport among the acidic units. As evidenced, the addition of phosphoric acid in Nafion yields super acids under an anhydrous state with the generation of excess $H_4PO_4^+$ and $-SO_3^-$ anions [67]. This reduces the proton transport through structural diffusion. To overcome the above problem, aromatic heterocyclic proton solvents are introduced in the PFSA matrix. Liquid solvents, such as triazole, imidazole, amino triazole, and benzimidazole, are introduced which serve the purpose of water in proton conduction process leading to improved mechanical properties and low volatility of protonic solvents. In this respect, poly(1-vinyl-1,2,4-triazole) (PVTri) is an interesting compound, introduction of which in Nafion as a blend results in improved proton conductivity when doped with phosphoric acid due to the presence of triazole repeating unit. Again, the high thermal and electrochemical stability of PVTri make it an attractive candidate for high-temperature fuel cell applications. Sen et al. have developed a blend membrane made of Nafion-PVTri at variable PVTri molar ratios [68]. The membranes exhibit high thermal stability up to 300°C. Further, the methanol permeability of these membranes is also lower than the Nafion 112 membrane. The proton conductivity attains a maximum of 10^{-3} S cm^{-1} when RH reaches 50% at ambient conditions. Polymer blending remains a common approach to prepare PEMs with improved physical and chemical properties. Sulfonated poly(vinylidene fluoride-co-hexafluoropropylene) (SPVdF-co-HFP) and sulfonated poly(ether sulfone) (SPES) have been also fabricated as blend membrane [69]. The AFM study of these membrane shows increased module size and surface roughness indicating the successful formation of the blend membrane (Figure 6.4). SPES is a suitable choice for the preparation of blend membrane with (SPVdF-co-HFP) because of its film-forming ability. The electrochemical performance, high thermal stability, considerable mechanical strength, high surface energy make them a suitable choice as PEM material. Again, SPES polymer contains polar groups along the polymer backbone that help in achieving high proton conductivity and water uptake properties in this polymer.

Other blend membranes such as PBI/polyether, PPO/poly(styrene-*b*-vinylbenzylphosphonic acid), SPEEK/PSU blends are also studied for fuel cell application [70–72]. Different solvent systems such as DMSO, DMAc, DMF, and NMP are used depending on the chemical nature and solubility of the polymer matrix to prepare homogeneous and compatible polymer films. The casted films are dried to obtain smooth films and finally acid-treated to induce the proton conduction property. Wang et al. have reported sulfonated poly[bis(benzimidazobenzisoquino linones)/poly(vinylidene fluoride) blend membranes by dissolving the polymers in

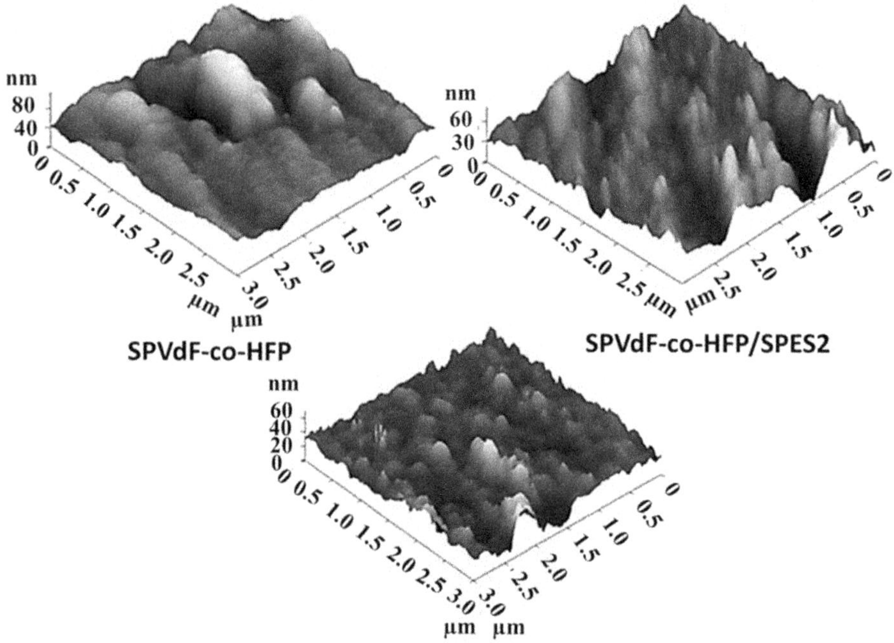

FIGURE 6.4 AFM images of SPVdF-co-HFP, SPVdF-co-HFP/SPES2 and SPVdF-co-HFP/SPES10 membranes [69](Reprinted with permission)

DMSO for 12 hrs. [73]. The homogeneous solution is filtered off through a PTFE membrane and cast into films.

The very first approach to prepare anhydrous proton-conducting membrane is the doping of polymers with phosphoric acid, sulfuric acid, etc. In this respect, phosphoric acid and heterocyclic compounds are the most common choices. Different polymers, like PBI, PVA, PEO, and poly(ethylenimine), are used as the host matrix materials, and the proton transport in this doped membrane is mainly governed by phosphoric acid unit by structural diffusion [74–77]. PVA is chosen as the common basic polymer with a good film-forming ability, remarkable thermal stability, and good mechanical strength. This water-soluble polymer easily reacts with different crosslinking agents to form a gel. The biocompatibility and biodegradability remain the major advantageous of this kind of system. The excellent selectivity of PVA for water to alcohol effectively reduces the methanol permeation through the membrane. Pure PVA does not show any ionic conduction property. For fuel cell applications, the proton conductivity is induced in PVA by sulfonating the polymer with different sulfonating agents, such as sulphuric acid, sulfosuccinic acid, and sulfophthalic acid. Anhydrous proton exchange membranes of cross-linked PVA and sulfosuccinic acid (SSA) are developed at varying SSA content. The blend membranes are prepared by mixing 3-amino-1,2,4-triazole (ATri) and 1H-1,2,4-triazole (Tri) at different weight ratios [78]. The membranes are characterized by excellent flexibility, improved thermal stability, and transparency. The methanol permeability of these membranes is also lower as compared to Nafion 112 membrane. The proton conductivity of the blend membranes is improved with an increase in ATri and $-SO_3H$ content due to the formation of ion transport channel and network structure in the blend membranes.

6.9 NATURAL POLYMER-BASED PEM

Polymer electrolytes based on environmentally friendly natural polymers are catching the recent interest. In this line, very few materials such as gelatin, pectin, chitin phosphate, agar, and alginic acid, are investigated as PEMs [79–85]. The research on cellulose-based materials is more extensive as compared to the other materials [86, 87]. Nanocrystalline cellulose is one of the most abundant natural polymers. The hydroxyl groups on nanocrystalline cellulose facilitate the hydrogen bond formation which, in turn, helps to transfer protons through the membrane. Moreover, these hydrogen bonds in the cellulose can interact with the sulfonic acid groups of sulfonated polyether ketones leading to the formation of membrane with excellent mechanical and dimensional stability. Hence the nanocomposite membranes consisting of nanocrystalline cellulose with sulfonated polyethers are expected to form efficient fuel cell membranes with desired proton conductivity. Wei et al. have fabricated a nanocomposite membrane using highly sulfonated polyether ether ketone ketone and sulfonated nanocrystalline cellulose through general solution blending and casting approach [88]. The incorporation of sulfonated nanocrystalline cellulose has been found to be advantageous in maintaining proper water content within the membrane as well as good mechanical properties. The nanocomposite membrane exhibits proton conductivity as high as 0.242 S cm^{-1} at 80°C. In a recent study, AMPS grafted cellulose PEMs have been reported to exhibit high proton conductivity, however, the power density of these membranes achieves 35% that of Nafion membranes under similar test conditions [89].

Chitosan is another natural polymer studied extensively as a natural polymer-based solid electrolyte membrane. It is a linear polysaccharide and naturally occurring cationic polyelectrolyte partially protonated with low intrinsic conductivity ($NH_2 + H_2O \leftrightharpoons NH_3 + OH^-$). The conduction is promoted by hydroxyl groups across the amorphous part of chitosan. Nonetheless, the intrinsic conductivity of chitosan is found to be around 10^{-4} S cm^{-1} which need to be further improved. The low proton conduction property of chitosan can be overcome by incorporating different inorganic filers and acids such as heteropolyacids, acidic oxides, ammonium salts, and grafting with acidic components[90–94]. The protons are conducted through the membrane via transport mechanism and Grotthuss pathways. Further, the addition of fillers reduces the mechanical strength of the membranes which can be addressed by introducing ionic or covalent crosslinks in chitosan [95]. Additionally, chitosan-based acid-base complex membranes can be used for high-temperature fuel cell applications as well. The conduction properties of hydrated and partially hydrated chitosan-based membranes are governed by mixed transport and Grotthuss mechanism. Figure 6.5 represents the anhydrous proton conduction process in chitosan.

Although chitosan-based membranes suffer from low proton conductivity value but they show significantly reduced methanol permeation properties which make them a material of choice for DMFC applications [96, 97]. Table 6.1 represents the comparison of properties of some PEMs superior to that of the standard Nafion membrane.

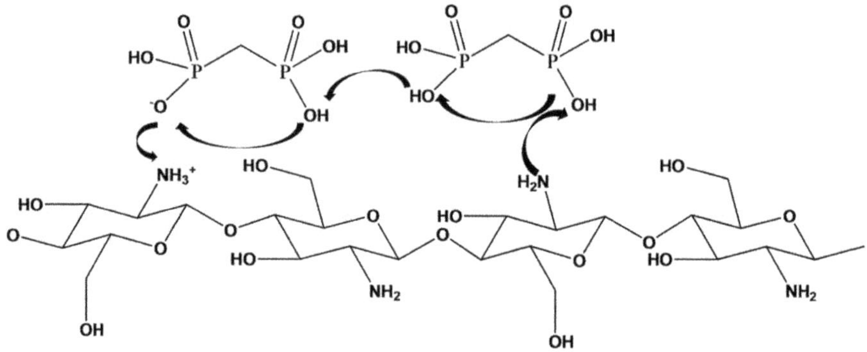

FIGURE 6.5 Proton conduction mechanism in anhydrous chitosan complex membranes

TABLE 6.1
Some systems with properties better than commercial Nafion®

Sl no.	Membrane	Proton conductivity (S cm^{-1})	Methanol permeability (cm^2 s^{-1})	Comments	References
1.	Fluorene-based poly(arylene ether sulfone)s	0.061–0.209 at 30°C and 0.146–0.365 at 80°C	3.22–13.1 × 10^{-7}	Nafion: 0.077 and 0.165 S cm^{-1} at 30 and 80°C, 15.5 × 10^{-7} cm^2 s^{-1}	[98]
2.	Nafion-Ceria composite membrane	176 × 10^{-3} at 70°C	-	Nafion: 136 × 10^{-3} at 70°C 30% higher than unmodified membrane	[99]
3.	Hybrid membranes of acid-base pair functionalized hollow mesoporous silica + Nafion	1.19 × 10^{-1} at 30°C and 100% RH	-	Conductivity 8 times upper than recast Nafion at 40°C and 20% RH after testing for 90 min	[100]
4.	Nafion and Fe_3O_4-anchored and Nafion-functionalized MWCNT	Proton conductivity 67–85 × 10^{-3} at 60°C and 95% RH	5.4 and 6.8× 10^{-7}	Recast Nafion: 31× 10^{-3} S cm^{-1} 30.1 × 10^{-7} cm^2 s^{-1}, Membrane selectivity 15 fold	[101]

(Continued)

TABLE 6.1
(Continued)

Sl no.	Membrane	Proton conductivity (S cm^{-1})	Methanol permeability (cm^2 s^{-1})	Comments	References
5.	Exfoliated Nafion–boron nitride nanocomposite	0.33 at 80°C and 95% RH	2.7–4.7× 10^{-8}	Recast Nafion: 1.8 × 10^{-8} cm^2 s^{-1}, Conductivity 6 times of recast Nafion	[102]
6.	Polynorbornene-based PEM	117.4 ± 3.7 × 10^{-3} at 25°C 213.2 ± 2.2 × 10^{-3} at 75°C.	1.0 ± 0.2 ×10^{-6}	Nafion 117: 86.6 ± 1.1 × 10^{-3} S/cm at 25°C, 154.7 ± 2.7 × 10^{-3} S/cm at 75°C, 2.1 ± 0.2×10^{-6} cm^2 s^{-1}	[103]
7.	Poly(arylene ether sulfone) with flexible acid side chains	0.108–0.258 and 0.135–0.194 at 80°C	1.59 × 10^{-7}–4.69 × 10^{-7}	Nafion: 0.077 and 0.165 S cm^{-1} at 30 and 80°C, 15.5 × 10^{-7} cm^2 s^{-1}	[104]
8.	Nafion nanohybrid membrane with phosphonic acid containing graphene oxide	0.277 at 100°C and100% RH 0.0441 at 80°C and 40% RH	-	Conductivity 1.2 and 6.6 times greater compared to pristine Nafion	[105]
9.	Sulfonated poly(arylene ether) membranes with triazole	0.211–0.256 at 80°C and 100% RH	-	Conductivity 1.5 times greater than Nafion 212 at 80°C	[106]

6.10 MICROCAPSULE-BASED PEM

With the urge to improve the proton conductivity of PEM under low relative humidity, microcapsules containing conducting and water retentive components are developed in recent days. In this respect, He et al. have studied modified PEM containing Zwitterionic microcapsules having sulfobutadiene [107]. The inclusion of these microcapsules by precipitation polymerization helps to improve the water retention and proton conduction ability of these membranes. The composite membrane containing 15 wt.% microcapsules exhibits proton conductivity of 5.8×10^{-2} S cm^{-1} under 20% RH at 40°C which is about 21 times better than the control Nafion membrane under similar test condition. The microcapsules act as the water reservoirs, which can assist conduct proton by providing additional proton conduction pathways inside the PEMs. In a similar work, phosphotungstic acid (HPW) loaded imidazole microcapsules have been incorporated into the SPEEK matrix as an acid reservoir [108]. The hollow imidazole microcapsules are absorbed in the HPW solution to prepare HPW-loaded imidazole microcapsules which in the subsequent process are embedded into the PEM to serve as the acid reservoir. The addition of HPW-filled imidazole microcapsules has significantly improved the proton conductivity of PEM and also exhibited reduced swelling and methanol permeation behavior. The composite membrane at a doping level of 15 wt.% has exhibited proton conductivity of about 3 times higher than the control SPEEK membrane. The insertion of HPW loaded microcapsules is helpful in retaining water inside the membrane and thereby providing additional proton conduction pathways to ease the conduction even at 20% RH. The composite membranes have exhibited two-fold better conductivity than the control SPEEK under 20% RH. With the similar aim to achieve better conductivity at low humidity and enhanced water retention capability, Li et al. have studied SPEEK composite membrane containing ellipsoidal polymeric microcapsules based on two different functional moieties, e.g., phosphoric acid and imidazole groups [109]. The water retention properties of composite membranes containing 5 wt.% phosphoric acid-filled microcapsules are three to nine-fold better than the pristine SPEEK membranes. The proton conductivity of composite membranes is observed to be quite impressive (3.51×10^{-3} S cm^{-1}) even under low RH and 60 minutes of test cycles. Zhang et al. have demonstrated a composite membrane made of ionic liquid-containing polymeric microcapsules embedded in the SPEEK matrix for operation under low relative humidity [110]. The composite membranes are fabricated by embedding methacrylic acid polymeric microcapsules into the SPEEK matrix which in the subsequent step have been impregnated in imidazole type ionic liquid. The microcapsules act as the reservoir of ionic liquid and permit heavy loading of ionic liquid into the membrane. High loading of ionic liquid helps to provide more hopping sites under anhydrous state where proton conduction proceed via the Grotthuss mechanism (Figure 6.6). At a loading of 12%, the proton conductivity reaches 33.7 mS cm^{-1} at 150°C with 36.9% retention of ionic liquid after performing leaching in water.

In another work, polymeric microcapsules bearing hydrophilic-hydrophobic properties have been embedded in the SPEEK matrix to improve the performance under low humidity [111]. The hydrophilicity of the microcapsules is controlled by regulating the content of hydrophilic ethylene glycol dimethacrylate (EGDMA) crosslinked with hydrophobic divinylbenzene (DVB). The microcapsules are capable of increasing proton conductivity up to thirteen times as compared to the control SPEEK membrane. The microcapsules also reduce the methanol permeability of the membrane and thereby requiring low activation energy for the proton conduction process. A comparative evaluation of proton conductivity of various membranes has been summarized for a broad overview of PEM and presented in Figure 6.7.

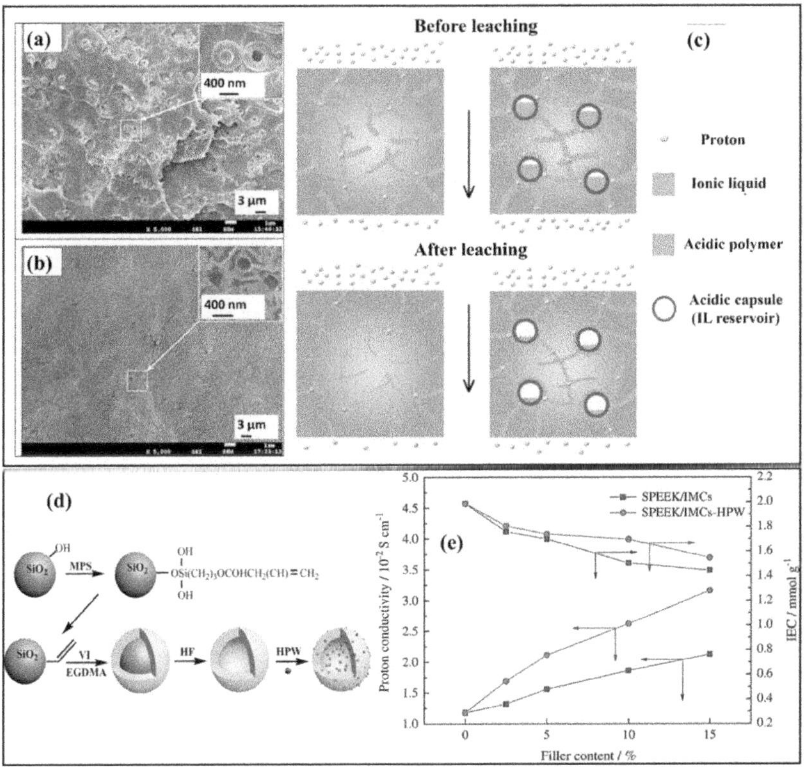

FIGURE 6.6 SEM images of SPEEK filled with (a) methacrylic acid polyelectrolyte microcapsules and (b) methacrylic acid polyelectrolyte microcapsules with ionic liquid, (c) schematic of proton transfer in the membranes before and after leaching of ionic liquid [110], (d) Schematic of HPW loaded in imidazole microcapsules, and (e) variation of proton conductivity of composite membranes loaded with microcapsules [108] (Reprinted with permission)

6.11 CHAPTER SUMMARY

In the quest of alternative PEM, research is focused on the development of low-cost, proton conductive polymer membranes. The alternative membranes explored are of several types such as hydrocarbon types, e.g., polysulfones, polyphosphazene, polyimides, polystyrene, microcapsule filled and natural polymer based membranes. The microcapsule based PEMs are fabricated by incorporating microcapsules filled with proton conductive ionic liquids that can contribute to the retention of proton conductivity during prolonged operation. The PEMs can also be prepared by blending of two polymers, first preferably containing hydrophilic groups contributing to the proton conductivity and second contributing to the mechanical integrity. The swelling tendency of the functionalized PEMs can be controlled by judicial crosslinking to obtain a balanced combination of water uptake capacity and dimensional stability. Natural polymers such as nanocrystalline cellulose and chitosan are also explored as polymer

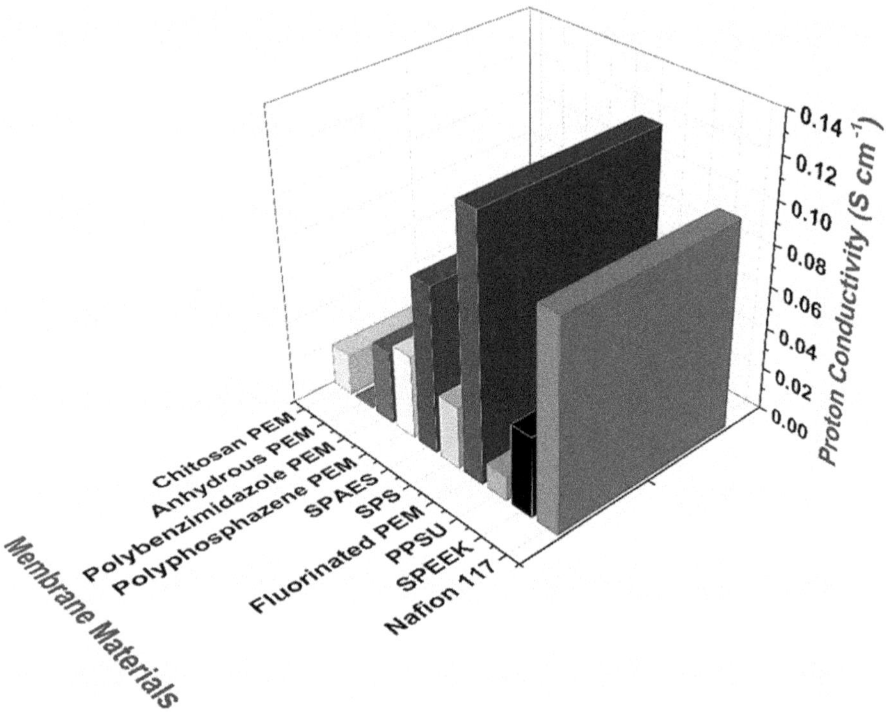

FIGURE 6.7 Proton conductivity of common membrane materials: Nafion 117 [112], sulfonated poly(ether ether ketone) (SPEEK) [113], phosphonated polysulfone (PPSU) [114], fluorinated PEM [115], sulfonated polystyrene (SPS) [116], sulfonated poly(aryl ether sulfone) (SPAES) [117], polyphosphazene PEM [11], polybenzimidazole PEM [118], anhydrous PEM (Nafion/azole) [119], and chitosan PEM [120]

electrolyte membranes. A comparative proton conductivity data of various alternative membranes has also been included based on the recent progress in this field.

REFERENCES

[1] Rikukawa M, Sanui K. Proton-conducting polymer electrolyte membranes based on hydrocarbon polymers. *Prog. Polym. Sci.* 25 2000:1463–502. https://doi.org/10.1016/S0079-6700(00)00032-0

[2] Shin DW, Guiver MD, Lee YM. Hydrocarbon-based polymer electrolyte membranes: importance of morphology on ion transport and membrane stability. *Chem Rev* 2017; 117:4759–805. https://doi.org/10.1021/acs.chemrev.6b00586.

[3] Lade H, Kumar V, Arthanareeswaran G, Ismail AF. Sulfonated poly(arylene ether sulfone) nanocomposite electrolyte membrane for fuel cell applications: A review. *Int J Hydrogen Energy* 2017;42:1063–74. https://doi.org/10.1016/j.ijhydene.2016.10.038.

[4] Singh B, Duong NMH, Henkensmeier D, Jang JH, Kim HJ, Han J, et al. Influence of different side-groups and cross-links on phosphoric acid doped radel-based polysulfone membranes for high temperature polymer electrolyte fuel cells. *Electrochim Acta* 2017;224:306–13. https://doi.org/10.1016/j.electacta.2016.12.088.

[5] Ozden A, Ercelik M, Devrim Y, Colpan CO, Hamdullahpur F. Evaluation of sulfonated polysulfone/zirconium hydrogen phosphate composite membranes for direct methanol fuel cells. *Electrochim Acta* 2017;256:196–210. https://doi.org/10.1016/j.electacta.2017.10.002.

[6] Kim K, Jung B-K, Ko T, Kim T-H, Lee J-C. Comb-shaped polysulfones containing sulfonated polytriazole side chains for proton exchange membranes. *J Memb Sci* 2018;554:232–43. https://doi.org/10.1016/j.memsci.2018.03.012.

[7] Krathumkhet N, Vongjitpimol K, Chuesutham T, Changkhamchom S, Phasuksom K, Sirivat A, et al. Preparation of sulfonated zeolite ZSM-5/sulfonated polysulfone composite membranes as PEM for direct methanol fuel cell application. *Solid State Ion* 2018;319:278–84. https://doi.org/10.1016/j.ssi.2018.02.019.

[8] Ahmadian-Alam L, Mahdavi H. A novel polysulfone-based ternary nanocomposite membrane consisting of metal-organic framework and silica nanoparticles: As proton exchange membrane for polymer electrolyte fuel cells. *Renew Energy* 2018;126:630–9. https://doi.org/10.1016/j.renene.2018.03.075.

[9] Kim J, Chung K, Lee H, Bae B, Cho E-B. Mesoporous ceria-silica/poly(arylene ether sulfone) composite membranes for durability of fuel cell electrolyte membrane. *Microporous Mesoporous Mater* 2016;236:292–300. https://doi.org/10.1016/j.micromeso.2016.01.049.

[10] Allcock HR, Fitzpatrick RJ, Salvati L. Sulfonation of (aryloxy)- and (arylamino) phosphazenes: Small-molecule compounds, polymers, and surfaces. *Chem Mater* 1991;3:1120–32. https://doi.org/10.1021/cm00018a032.

[11] Guo Q, N. Pintauro P, Tang H, O'Connor S. Sulfonated and crosslinked polyphosphazene-based proton-exchange membranes. *J Memb Sci* 1999;154:175–81. https://doi.org/10.1016/S0376-7388(98)00282-8.

[12] Zhao Z, Zhou Z, Jianbing JI, Xie F. Electrochemical properties of poly(aminobenzene sulfonic)phosphazene derivative grafted SPEEK proton exchange membrane. *Int J Electrochem Sci* 2015;10:4303–13. https://doi.org/10.1016/S1452-3981(23)06623-3.

[13] Dong Y, Zhang Y, Xu H, Luo T, Fu F, Zhu C. Proton exchange membrane based on crosslinked sulfonated polyphosphazene containing pendent perfluorosulfonic acid groups with sulfonated poly(ether ether ketone). *J Appl Polym Sci* 2016;133:43492–500. https://doi.org/10.1002/app.43492.

[14] Tang H, Pintauro PN. Polyphosphazene membranes. IV. Polymer morphology and proton conductivity in sulfonated poly[bis(3-methylphenoxy)phosphazene] films. *J Appl Polym Sci* 2001;79:49–59. https://doi.org/10.1002/1097-4628(20010103)79:1<49::AID-APP60>3.0.CO;2-J.

[15] Allcock HR, Wood RM. Design and synthesis of ion-conductive polyphosphazenes for fuel cell applications: Review. *J Polym Sci B Polym Phys* 2006;44:2358–68. https://doi.org/10.1002/polb.20864.

[16] Zhou X, Weston J, Chalkova E, Hofmann MA, Ambler CM, Allcock HR, et al. High temperature transport properties of polyphosphazene membranes for direct methanol fuel cells. *Electrochim Acta* 2003;48:2173–80. https://doi.org/10.1016/S0013-4686(03)00201-9.

[17] Koppel IA, Taft RW, Anvia F, Zhu S-Z, Hu L-Q, Sung K-S, et al. The gas-phase acidities of very strong neutral bronsted acids. *J Am Chem Soc* 1994;116:3047–57. https://doi.org/10.1021/ja00086a038.

[18] Hofmann MA, Ambler CM, Maher AE, Chalkova E, Zhou XY, Lvov SN, et al. Synthesis of polyphosphazenes with sulfonimide side groups. *Macromolecules* 2002;35:6490–3. https://doi.org/10.1021/ma020330z.

[19] Rodgers M, Yang Y, Holdcroft S. A study of linear versus angled rigid rod polymers for proton conducting membranes using sulfonated polyimides. *Eur Polym J* 2006;42:1075–85. https://doi.org/10.1016/j.eurpolymj.2005.11.011.

[20] Cornet N, Diat Olivier, Gebel G, Jousse F, Marsacq D, Mercier Regis, et al. Sulfonated polyimide membranes: A new type of ion-conducting membrane for electrochemical applications. *J New Mater Electrochem Syst* 2000;3:33–42.

[21] Vallejo E, Pourcelly G, Gavach C, Mercier R, Pineri M. Sulfonated polyimides as proton conductor exchange membranes. Physicochemical properties and separation H+/Mz+ by electrodialysis comparison with a perfluorosulfonic membrane. *J Memb Sci* 1999;160:127–37. https://doi.org/10.1016/S0376-7388(99)00070-8.

[22] Woo Y, Oh SY, Kang YS, Jung B. Synthesis and characterization of sulfonated polyimide membranes for direct methanol fuel cell. *J Memb Sci* 2003;220:31–45. https://doi.org/10.1016/S0376-7388(03)00185-6.

[23] He Y, Tong C, Geng L, Liu L, Lü C. Enhanced performance of the sulfonated polyimide proton exchange membranes by graphene oxide: Size effect of graphene oxide. *J Memb Sci* 2014;458:36–46. https://doi.org/10.1016/j.memsci.2014.01.017.

[24] Miyatake K, Zhou H, Uchida H, Watanabe M. Highly proton conductive polyimide electrolytes containing fluorenyl groupsEleupplementary information (ESI) available: Experimental section. *Chem Commun* 2003:368–9. https://doi.org/10.1039/b210296j. See https://www.rsc.org/suppdata/cc/b2/b210296j/.

[25] Chen X, Chen K, Chen P, Higa M, Okamoto K, Hirano T. Effects of tetracarboxylic dianhydrides on the properties of sulfonated polyimides. *J Polym Sci A Polym Chem* 2010;48:905–15. https://doi.org/10.1002/pola.23844.

[26] Genies C, Mercier R, Sillion B, Cornet N, Gebel G, Pineri M. Soluble sulfonated naphthalenic polyimides as materials for proton exchange membranes. *Polymer* 2001;42:359–73. https://doi.org/10.1016/S0032-3861(00)00384-0.

[27] Savadogo O. Emerging membrane for electrochemical systems:(I) solid polymer electrolyte membranes for fuel cell systems. *J New Mater Electrochem Syst* 1998;1:47–66.

[28] Asano N, Aoki M, Suzuki S, Miyatake K, Uchida H, Watanabe M. Aliphatic/aromatic polyimide ionomers as a proton conductive membrane for fuel cell applications. *J Am Chem Soc* 2006;128:1762–9. https://doi.org/10.1021/ja0571491.

[29] Asano N, Miyatake K, Watanabe M. Hydrolytically stable polyimide ionomer for fuel cell applications. *Chem Mater* 2004;16:2841–3. https://doi.org/10.1021/cm049550c.

[30] Wei H, Fang X. Novel aromatic polyimide ionomers for proton exchange membranes: Enhancing the hydrolytic stability. *Polymer* 2011;52:2735–9. https://doi.org/10.1016/j.polymer.2011.04.047.

[31] Zhang F, Li N, Cui Z, Zhang S, Li S. Novel acid–base polyimides synthesized from binaphthalene dianhydrie and triphenylamine-containing diamine as proton exchange membranes. *J Memb Sci* 2008;314:24–32. https://doi.org/10.1016/j.memsci.2008.01.032.

[32] Zhang F, Li N, Zhang S, Li S. Ionomers based on multisulfonated perylene dianhydride: Synthesis and properties of water resistant sulfonated polyimides. *J Power Sources* 2010;195:2159–65. https://doi.org/10.1016/j.jpowsour.2009.10.026.

[33] Belkhiri Z, Zeroual M, Ben Moussa H, Zitouni B. Effect of temperature and water content on the performance of PEM fuel cell. *J Renew Energ* 2023;14. https://doi.org/10.54966/jreen.v14i1.246.

[34] Ahmed M, Dincer I. A review on methanol crossover in direct methanol fuel cells: Challenges and achievements. *Int J Energy Res* 2011;35:1213–28. https://doi.org/10.1002/er.1889.

[35] Dockheer SM, Gubler L, Bounds PL, Domazou AS, Scherer GG, Wokaun A, et al. Damage to fuel cell membranes. Reaction of HO˙ with an oligomer of poly(sodium styrene sulfonate) and subsequent reaction with O2. *Phys Chem Chem Phys* 2010;12:11609–16. https://doi.org/10.1039/c0cp00082e.

[36] Assink RA, Arnold C, Hollandsworth RP. Preparation of oxidatively stable cation-exchange membranes by the elimination of tertiary hydrogens. *J Memb Sci* 1991;56:143–51. https://doi.org/10.1016/S0376-7388(00)80804-2.

[37] Elabd YA, Napadensky E, Sloan JM, Crawford DM, Walker CW. Triblock copolymer ionomer membranes. *J Memb Sci* 2003;217:227–42. https://doi.org/10.1016/S0376-7388(03)00127-3.
[38] Elabd YA, Napadensky E, Walker CW, Winey KI. Transport properties of sulfonated poly(styrene- *b* -isobutylene- *b* -styrene) triblock copolymers at high ion-exchange capacities. *Macromolecules* 2006;39:399–407. https://doi.org/10.1021/ma051958n.
[39] Gubler L, Gürsel SA, Scherer GG. Radiation grafted membranes for polymer electrolyte fuel cells. *Fuel Cells* 2005;5:317–35. https://doi.org/10.1002/fuce.200400078.
[40] Enomoto K, Takahashi S, Rohani R, Maekawa Y. Synthesis of copolymer grafts containing sulfoalkyl and hydrophilic groups in polymer electrolyte membranes. *J Memb Sci* 2012;415–416:36–41. https://doi.org/10.1016/j.memsci.2012.04.035.
[41] Chen J, Septiani U, Asano M, Maekawa Y, Kubota H, Yoshida M. Comparative study on the preparation and properties of radiation-grafted polymer electrolyte membranes based on fluoropolymer films. *J Appl Polym Sci* 2007;103:1966–72. https://doi.org/10.1002/app.24950.
[42] Gubler L. Polymer design strategies for radiation-grafted fuel cell membranes. *Adv Energy Mater* 2014;4:1300827–50. https://doi.org/10.1002/aenm.201300827.
[43] Holmberg S, Lehtinen T, Näsman J, Ostrovskii D, Paronen M, Serimaa R, et al. Structure and properties of sulfonated poly [(vinylidene fluoride)–g-styrene] norous membranes porous membranes. *J Mater Chem* 1996;6:1309–17. https://doi.org/10.1039/JM9960601309.
[44] Gubler L, Slaski M, Wallasch F, Wokaun A, Scherer GG. Radiation grafted fuel cell membranes based on co-grafting of α-methylstyrene and methacrylonitrile into a fluoropolymer base film. *J Memb Sci* 2009;339:68–77. https://doi.org/10.1016/j.memsci.2009.04.031.
[45] Kumar P, Dutta K, Das S, Kundu PP. Membrane prepared by incorporation of crosslinked sulfonated polystyrene in the blend of PVdF-co-HFP/Nafion: A preliminary evaluation for application in DMFC. *Appl Energy* 2014;123:66–74. https://doi.org/10.1016/j.apenergy.2014.02.060.
[46] Kumar P, Jagwani SK, Kundu PP. A study on the heat behaviour of PEM, prepared by incorporation of crosslinked sulfonated polystyrene in the blend of PVdF-co-HFP/Nafion, for its high temperature application in DMFC. *Mater Today Commun* 2015;2:e1–8. https://doi.org/10.1016/j.mtcomm.2014.10.002.
[47] Isaacs Sodeye AI, Huang T, Gido SP, Mays JW. Polymer electrolyte membranes from fluorinated polyisoprene-block-sulfonated polystyrene: Membrane structure and transport properties. *Polymer* 2011;52:1963–70. https://doi.org/10.1016/j.polymer.2011.02.049.
[48] Bae B, Kim D. Sulfonated polystyrene grafted polypropylene composite electrolyte membranes for direct methanol fuel cells. *J Memb Sci* 2003;220:75–87. https://doi.org/10.1016/S0376-7388(03)00216-3.
[49] Baglio V, Stassi A, Modica E, Antonucci V, Aricò AS, Caracino P, et al. Performance comparison of portable direct methanol fuel cell mini-stacks based on a low-cost fluorine-free polymer electrolyte and Nafion membrane. *Electrochim Acta* 2010;55:6022–7. https://doi.org/10.1016/j.electacta.2010.05.059.
[50] Di Y, Yang W, Li X, Zhao Z, Wang M, Dai J. Preparation and characterization of continuous carbon nanofiber-supported SPEEK composite membranes for fuel cell application. *RSC Adv* 2014;4:52001–7. https://doi.org/10.1039/C4RA06976E.
[51] Liu X, Yang Z, Zhang Y, Li C, Dong J, Liu Y, et al. Electrospun multifunctional sulfonated carbon nanofibers for design and fabrication of SPEEK composite proton exchange membranes for direct methanol fuel cell application. *Int J Hydrogen Energy* 2017;42:10275–84. https://doi.org/10.1016/j.ijhydene.2017.02.128.
[52] Zhnag Shu. *Mechanical and physical properties of electrospun nanofibers.* North Carolina State University; 2009.

[53] Lee K-S, Jeong M-H, Lee J-P, Lee J-S. End-group cross-linked poly(arylene ether) for proton exchange membranes. *Macromolecules* 2009;42:584–90. https://doi.org/10.1021/ma802233j.

[54] Sun F, Qin L-L, Zhou J, Wang Y-K, Rong J-Q, Chen Y-J, et al. Friedel-Crafts self-cross-linking of sulfonated poly(etheretherketone) composite proton exchange membrane doped with phosphotungstic acid and carbon-based nanomaterials for fuel cell applications. *J Memb Sci* 2020;611:118381. https://doi.org/10.1016/j.memsci.2020.118381.

[55] Liu C-P, Dai C-A, Chao C-Y, Chang S-J. Novel proton exchange membrane based on crosslinked poly(vinyl alcohol) for direct methanol fuel cells. *J Power Sources* 2014;249:285–98. https://doi.org/10.1016/j.jpowsour.2013.10.117.

[56] Hasiotis C, Deimede V, Kontoyannis C. New polymer electrolytes based on blends of sulfonated polysulfones with polybenzimidazole. *Electrochim Acta* 2001;46:2401–6. https://doi.org/10.1016/S0013-4686(01)00437-6.

[57] Bouzek K, Moravcová S, Samec Z, Schauer J. H[sup +] and Na[sup +] Ion transport properties of sulfonated poly(2,6-dimethyl-1,4-phenyleneoxide) membranes. *J Electrochem Soc* 2003;150:E329–36. https://doi.org/10.1149/1.1572484.

[58] Hong YT, Lee CH, Park HS, Min KA, Kim HJ, Nam SY, et al. Improvement of electrochemical performances of sulfonated poly(arylene ether sulfone) via incorporation of sulfonated poly(arylene ether benzimidazole). *J Power Sources* 2008;175:724–31. https://doi.org/10.1016/j.jpowsour.2007.09.068.

[59] Wycisk R, Pintauro PN, Wang W, O'Connor S. Polyphosphazene membranes. I. Solid-state photocrosslinking of poly[(4-ethylphenoxy)-(phenoxy)phosphazene]. *J Appl Polym Sci* 1996;59:1607–17. https://doi.org/10.1002/(SICI)1097-4628(19960307)59:10<1607::AID-APP13>3.0.CO;2-T.

[60] Mikhailenko SD, Wang K, Kaliaguine S, Xing P, Robertson GP, Guiver MD. Proton conducting membranes based on cross-linked sulfonated poly(ether ether ketone) (SPEEK). *J Memb Sci* 2004;233:93–9. https://doi.org/10.1016/j.memsci.2004.01.004.

[61] Lee CH, Park HB, Chung YS, Lee YM, Freeman BD. Water sorption, proton conduction, and methanol permeation properties of sulfonated polyimide membranes cross-linked with *N*, *N* - Bis(2-hydroxyethyl)-2-aminoethanesulfonic Acid (BES). *Macromolecules* 2006;39:755–64. https://doi.org/10.1021/ma052226y.

[62] Rhim J, Park H, Lee C, Jun J, Kim D, Lee Y. Crosslinked poly(vinyl alcohol) membranes containing sulfonic acid group: Proton and methanol transport through membranes. *J Memb Sci* 2004;238:143–51. https://doi.org/10.1016/j.memsci.2004.03.030.

[63] Jeong M-H, Lee K-S, Lee J-S. Cross-linking density effect of fluorinated aromatic polyethers on transport properties. *Macromolecules* 2009;42:1652–8. https://doi.org/10.1021/ma8024199.

[64] Di Vona ML, Sgreccia E, Tamilvanan M, Khadhraoui M, Chassigneux C, Knauth P. High ionic exchange capacity polyphenylsulfone (SPPSU) and polyethersulfone (SPES) cross-linked by annealing treatment: Thermal stability, hydration level and mechanical properties. *J Memb Sci* 2010;354:134–41. https://doi.org/10.1016/j.memsci.2010.02.058.

[65] He R. Proton conductivity of phosphoric acid doped polybenzimidazole and its composites with inorganic proton conductors. *J Memb Sci* 2003;226:169–84. https://doi.org/10.1016/j.memsci.2003.09.002.

[66] Pu H, Meyer WH, Wegner G. Proton transport in polybenzimidazole blended with H_3PO_4 or H_2SO_4. *J Polym Sci B Polym Phys* 2002;40:663–9. https://doi.org/10.1002/polb.10132.

[67] Savinell R, Yeager E, Tryk D, Landau U, Wainright J, Weng D, et al. A polymer electrolyte for operation at temperatures up to 200°C. *J Electrochem Soc* 1994;141:L46–8. https://doi.org/10.1149/1.2054875.

[68] Sen U, Bozkurt A, Ata A. Nafion/poly(1-vinyl-1,2,4-triazole) blends as proton conducting membranes for polymer electrolyte membrane fuel cells. *J Power Sources* 2010;195:7720–6. https://doi.org/10.1016/j.jpowsour.2010.04.087.

[69] Uma Devi A, Muthumeenal A, Sabarathinam RM, Nagendran A. Fabrication and electrochemical properties of SPVdF-co-HFP/SPES blend proton exchange membranes for direct methanol fuel cells. *Renew Energy* 2017;102:258–65. https://doi.org/10.1016/j.renene.2016.10.060.

[70] Cho CG, Kim SH, Park YC, Kim H, Park J-W. Fuel cell membranes based on blends of PPO with poly(styrene-b-vinylbenzylphosphonic acid) copolymers. *J Memb Sci* 2008;308:96–106. https://doi.org/10.1016/j.memsci.2007.09.052.

[71] Fu Y, Manthiram A, Guiver MD. Blend membranes based on sulfonated poly(ether ether ketone) and polysulfone bearing benzimidazole side groups for proton exchange membrane fuel cells. *Electrochem Commun* 2006;8:1386–90. https://doi.org/10.1016/j.elecom.2006.06.018.

[72] Li QF, Rudbeck HC, Chromik A, Jensen JO, Pan C, Steenberg T, et al. Properties, degradation and high temperature fuel cell test of different types of PBI and PBI blend membranes. *J Memb Sci* 2010;347:260–70. https://doi.org/10.1016/j.memsci.2009.10.032.

[73] Wang J, Li N, Cui Z, Zhang S, Xing W. Blends based on sulfonated poly[bis(benzi midazobenzisoquinolinones)] and poly(vinylidene fluoride) for polymer electrolyte membrane fuel cell. *J Memb Sci* 2009;341:155–62. https://doi.org/10.1016/j.memsci.2009.06.002.

[74] Bose S, Kuila T, Nguyen TXH, Kim NH, Lau K, Lee JH. Polymer membranes for high temperature proton exchange membrane fuel cell: Recent advances and challenges. *Prog Polym Sci* 2011;36:813–43. https://doi.org/10.1016/j.progpolymsci.2011.01.003.

[75] DONOSO P. NMR, conductivity and neutron scattering investigation of ionic dynamics in the anhydrous polymer protonic conductor PEO(H3PO4)x). *Solid State Ion* 1988;28–30:969–74. https://doi.org/10.1016/0167-2738(88)90313-X.

[76] Petty-Weeks S. Proton conducting interpenetrating polymer networks. *Solid State Ion* 1988;31:117–25. https://doi.org/10.1016/0167-2738(88)90295-0.

[77] Tanaka R, Yamamoto H, Kawamura S, Iwase T. Proton conducting behavior of poly(ethylenimine)-H3PO4 systems. *Electrochim Acta* 1995;40:2421–4. https://doi.org/10.1016/0013-4686(95)00207-U.

[78] Boroglu MS, Celik SU, Boz I, Bozkurt A. Sulfonated poly(vinyl alcohol)/triazole blends as anhydrous proton conducting membranes for polymer electrolyte membrane fuel cells. *J Mater Res* 2013;28:1458–65. https://doi.org/10.1557/jmr.2013.111.

[79] Honma I, Yamada M. Bio-inspired membranes for advanced polymer electrolyte fuel cells. anhydrous proton-conducting membrane via molecular self-assembly. *Bull Chem Soc Jpn* 2007;80:2110–23. https://doi.org/10.1246/bcsj.80.2110.

[80] Mattos RI, Pawlicka A, Lima JF, Tambelli CE, Magon CJ, Donoso JP. Magnetic resonance and conductivity study of gelatin-based proton conductor polymer electrolytes. *Electrochim Acta* 2010;55:1396–400. https://doi.org/10.1016/j.electacta.2009.04.038.

[81] Mishra RK, Datt M, Banthia AK, Majeed ABA. Development of novel pectin based membranes as proton conducting material. *Int J Plast Technol* 2012;16:80–8. https://doi.org/10.1007/s12588-012-9031-7.

[82] Pawlicka A, Mattos RI, Lima JF, Tambelli CE, Magon CJ, Donoso JP. Magnetic resonance and conductivity study of a gelatin-based polymer gel electrolyte. *Electrochim Acta* 2011;57:187–91. https://doi.org/10.1016/j.electacta.2011.07.062.

[83] Raphael E, Avellaneda CO, Manzolli B, Pawlicka A. Agar-based films for application as polymer electrolytes. *Electrochim Acta* 2010;55:1455–9. https://doi.org/10.1016/j.electacta.2009.06.010.

[84] Yamada M, Honma I. An anhydrous proton conductor based on lactam–lactim tautomerism of uracil. *ChemPhysChem* 2004;5:724–8. https://doi.org/10.1002/cphc.200301015.

[85] Yamada M, Honma I. Biomembranes for fuel cell electrolytes employing anhydrous proton conducting uracil composites. *Biosens Bioelectron* 2006;21:2064–9. https://doi.org/10.1016/j.bios.2006.01.022.

[86] Harun NI, Ali RM, Ali AMM, Yahya MZA. Dielectric behaviour of cellulose acetate-based polymer electrolytes. *Ionics (Kiel)* 2012;18:599–606. https://doi.org/10.1007/s11581-011-0653-0.

[87] Lopes LVS, Machado GO, Pawlicka A, Donoso JP. Nuclear magnetic resonance and conductivity study of hydroxyethylcellulose based polymer gel electrolytes. *Electrochim Acta* 2005;50:3978–84. https://doi.org/10.1016/j.electacta.2005.02.056.

[88] Wei Y, Shang Y, Ni C, Zhang H, Li X, Liu B, et al. Modified nanocrystal cellulose/fluorene-containing sulfonated poly(ether ether ketone ketone) composites for proton exchange membranes. *Appl Surf Sci* 2017;416:996–1006. https://doi.org/10.1016/j.apsusc.2017.04.190.

[89] Lin CW, Liang SS, Chen SW, Lai JT. Sorption and transport properties of 2-acrylamido-2-methyl-1-propanesulfonic acid-grafted bacterial cellulose membranes for fuel cell application. *J Power Sources* 2013;232:297–305. https://doi.org/10.1016/j.jpowsour.2013.01.047.

[90] Du JF, Bai Y, Pan DA, Chu WY, Qiao LJ. Characteristics of proton conducting polymer electrolyte based on chitosan acetate complexed with CH_3 $COONH_4$. *J Polym Sci B Polym Phys* 2009;47:549–54. https://doi.org/10.1002/polb.21656.

[91] Hassan F, Woo HJ, Aziz NA, Kufian MZ, Majid SR. Synthesis of Al2TiO5 and its effect on the properties of chitosan–NH4SCN polymer electrolytes. *Ionics (Kiel)* 2013;19:483–9. https://doi.org/10.1007/s11581-012-0763-3.

[92] Ramírez-Salgado J. Study of basic biopolymer as proton membrane for fuel cell systems. *Electrochim Acta* 2007;52:3766–78. https://doi.org/10.1016/j.electacta.2006.10.051.

[93] Xiang Y, Yang M, Guo Z, Cui Z. Alternatively chitosan sulfate blending membrane as methanol-blocking polymer electrolyte membrane for direct methanol fuel cell. *J Memb Sci* 2009;337:318–23. https://doi.org/10.1016/j.memsci.2009.04.006.

[94] Xiao Y, Xiang Y, Xiu R, Lu S. Development of cesium phosphotungstate salt and chitosan composite membrane for direct methanol fuel cells. *Carbohydr Polym* 2013;98:233–40. https://doi.org/10.1016/j.carbpol.2013.06.017.

[95] Mittal V, editor. *Polymers for energy storage and conversion*. Wiley; 2013. https://doi.org/10.1002/9781118734162.

[96] Gümüşoğlu T, Arı GA, Deligöz H. Investigation of salt addition and acid treatment effects on the transport properties of ionically cross-linked polyelectrolyte complex membranes based on chitosan and polyacrylic acid. *J Memb Sci* 2011;376:25–34. https://doi.org/10.1016/j.memsci.2011.03.040.

[97] Ma J, Sahai Y. Chitosan biopolymer for fuel cell applications. *Carbohydr Polym* 2013;92:955–75. https://doi.org/10.1016/j.carbpol.2012.10.015.

[98] Wang C, Li N, Shin DW, Lee SY, Kang NR, Lee YM, et al. Fluorene-based poly(arylene ether sulfone)s containing clustered flexible pendant sulfonic acids as proton exchange membranes. *Macromolecules* 2011;44:7296–306. https://doi.org/10.1021/ma2015968.

[99] Velayutham P, Sahu A, Parthasarathy S. A nafion-ceria composite membrane electrolyte for reduced methanol crossover in direct methanol fuel cells. *Energies (Basel)* 2017;10:259–71. https://doi.org/10.3390/en10020259.

[100] Yin Y, Deng W, Wang H, Li A, Wang C, Jiang Z, et al. Fabrication of hybrid membranes by incorporating acid–base pair functionalized hollow mesoporous silica for enhanced proton conductivity. *J Mater Chem A Mater* 2015;3:16079–88. https://doi.org/10.1039/C5TA03276H.

[101] Chang C-M, Li H-Y, Lai J-Y, Liu Y-L. Nanocomposite membranes of Nafion and Fe3O4-anchored and Nafion-functionalized multiwalled carbon nanotubes exhibiting high proton conductivity and low methanol permeability for direct methanol fuel cells. *RSC Adv* 2013;3:12895–904. https://doi.org/10.1039/c3ra40438b.

[102] Jia W, Tang B, Wu P. Novel composite proton exchange membrane with connected long-range ionic nanochannels constructed via exfoliated nafion–boron nitride nanocomposite. *ACS Appl Mater Interfaces* 2017;9:14791–800. https://doi.org/10.1021/acsami.7b00858.
[103] Li X, Zhao Y, Feng Z, Xiang X, Wang S, Xie X, et al. Ring-opening metathesis polymerization for the preparation of polynorbornene-based proton exchange membranes with high proton conductivity. *J Memb Sci* 2017;528:55–63. https://doi.org/10.1016/j.memsci.2016.12.050.
[104] Wang C, Shin DW, Lee SY, Kang NR, Lee YM, Guiver MD. Poly(arylene ether sulfone) proton exchange membranes with flexible acid side chains. *J Memb Sci* 2012;405–406:68–78. https://doi.org/10.1016/j.memsci.2012.02.045.
[105] Zhang B, Cao Y, Jiang S, Li Z, He G, Wu H. Enhanced proton conductivity of Nafion nanohybrid membrane incorporated with phosphonic acid functionalized graphene oxide at elevated temperature and low humidity. *J Memb Sci* 2016;518:243–53. https://doi.org/10.1016/j.memsci.2016.07.032.
[106] Ahn M-K, Lee S-B, Min C-M, Yu Y-G, Jang J, Gim M-Y, et al. Enhanced proton conductivity at low humidity of proton exchange membranes with triazole moieties in the side chains. *J Memb Sci* 2017;523:480–6. https://doi.org/10.1016/j.memsci.2016.10.018.
[107] He G, Li Z, Li Y, Li Z, Wu H, Yang X, et al. Zwitterionic Microcapsules as Water Reservoirs and Proton Carriers within a Nafion Membrane To Confer High Proton Conductivity under Low Humidity. *ACS Appl Mater Interfaces* 2014;6:5362–6. https://doi.org/10.1021/am500626f.
[108] Wu H, Shen X, Cao Y, Li Z, Jiang Z. Composite proton conductive membranes composed of sulfonated poly(ether ether ketone) and phosphotungstic acid-loaded imidazole microcapsules as acid reservoirs. *J Memb Sci* 2014;451:74–84. https://doi.org/10.1016/j.memsci.2013.09.058.
[109] Li Z, Yue X, He G, Li Z, Yin Y, Gang M, et al. Enhanced water retention and low-humidity proton conductivity of sulfonated poly(ether ether ketone) hybrid membrane by incorporating ellipsoidal microcapsules. *Int J Hydrogen Energy* 2015;40:8398–406. https://doi.org/10.1016/j.ijhydene.2015.04.138.
[110] Zhang H, Wu W, Li Y, Liu Y, Wang J, Zhang B, et al. Polyelectrolyte microcapsules as ionic liquid reservoirs within ionomer membrane to confer high anhydrous proton conductivity. *J Power Sources* 2015;279:667–77. https://doi.org/10.1016/j.jpowsour.2015.01.066.
[111] He G, Li Y, Li Z, Nie L, Wu H, Yang X, et al. Enhancing water retention and low-humidity proton conductivity of sulfonated poly(ether ether ketone) composite membrane enabled by the polymer-microcapsules with controllable hydrophilicity–hydrophobicity. *J Power Sources* 2014;248:951–61. https://doi.org/10.1016/j.jpowsour.2013.10.017.
[112] www.fuelcellstore.com. Accessed on 1st June, 2020 n.d.
[113] Xing P, Robertson GP, Guiver MD, Mikhailenko SD, Wang K, Kaliaguine S. Synthesis and characterization of sulfonated poly(ether ether ketone) for proton exchange membranes. *J Memb Sci* 2004;229:95–106. https://doi.org/10.1016/j.memsci.2003.09.019.
[114] Abu-Thabit NY, Ali SA, Javaid Zaidi SM. New highly phosphonated polysulfone membranes for PEM fuel cells. *J Memb Sci* 2010;360:26–33. https://doi.org/10.1016/j.memsci.2010.04.041.
[115] Norsten TB, Guiver MD, Murphy J, Astill T, Navessin T, Holdcroft S, et al. Highly fluorinated comb-shaped copolymers as proton exchange membranes (PEMs): Improving PEM properties through rational design. *Adv Funct Mater* 2006;16:1814–22. https://doi.org/10.1002/adfm.200500763.

[116] Kim J, Kim B, Jung B. Proton conductivities and methanol permeabilities of membranes made from partially sulfonated polystyrene-block-poly(ethylene-ran-butylene)-block-polystyrene copolymers. *J Memb Sci* 2002;207:129–37. https://doi.org/10.1016/S0376-7388(02)00138-2.

[117] Wang F, Hickner M, Kim YS, Zawodzinski TA, McGrath JE. Direct polymerization of sulfonated poly(arylene ether sulfone) random (statistical) copolymers: Candidates for new proton exchange membranes. *J Memb Sci* 2002;197:231–42. https://doi.org/10.1016/S0376-7388(01)00620-2.

[118] Asensio JA, Borrós S, Gómez-Romero P. Enhanced conductivity in polyanion-containing polybenzimidazoles. Improved materials for proton-exchange membranes and PEM fuel cells. *Electrochem Commun* 2003;5:967–72. https://doi.org/10.1016/j.elecom.2003.09.007.

[119] Sen U, Unugurcelik S, Ata A, Bozkurt A. Anhydrous proton conducting membranes for PEM fuel cells based on Nafion/Azole composites. *Int J Hydrogen Energy* 2008;33:2808–15. https://doi.org/10.1016/j.ijhydene.2008.03.007.

[120] Mukoma P, Jooste BR, Vosloo HCM. Synthesis and characterization of cross-linked chitosan membranes for application as alternative proton exchange membrane materials in fuel cells. *J Power Sources* 2004;136:16–23. https://doi.org/10.1016/j.jpowsour.2004.05.027.

7 Characterization of PEM and performance evaluation of PEMFC

7.1 MEMBRANE PROPERTIES

7.1.1 Ion exchange capacity

For evaluation of membrane properties, determination of the ion exchange capacity (IEC) of the membranes is very important. It is a very crucial parameter which indicates the ability of the membranes to conduct protons. IEC is defined as the number of protons that can be exchanged per unit mass of the membrane and is generally expressed as miliequivalent or milimoles per gram of dry membrane. A high value of IEC of the proton conducting membrane indicates good performance during fuel cell operation. IEC of the polymeric membranes are evaluated by titrimetric method using Eqn. 7.1. The membranes are first immersed overnight in NaCl solution to substitute entire H^+ ions with Na^+ ions and the protons freed by this process are back titrated with standard NaOH solution.

$$\mathrm{IEC} = (\text{Consumed NaOH in mL} \times \text{Molarity of NaOH})/W_{\mathrm{dry}} \tag{7.1}$$

The commercial Nafion membrane exhibits IEC in the range of 0.7–1.15 mmol/g depending on its grade as described in the literature [1]. The structural arrangement and composite formation have a significant impact on the IEC value of the PEMs. Sulfonation or introduction of any hydrophilic and proton conducting groups which contribute to the additional protons in the medium will improve the IEC of membrane [2]. Moreover, the incorporation of proton conductors also plays a dominant role in increasing the IEC of the membrane [2]. On the contrary, formation of crosslinks between the sulfonic or carboxylic acid functional groups reduce the overall acidity of the membranes and hence a reduction in IEC value is observed with the increase in crosslinking due to the reduction in numbers of acidic sites in the membranes as well as restriction in ionic mobility. This observation is common in the case of SPPEK/PBI membranes where there is a possibility of formation of crosslinked structure between the sulfonic acid groups of SPEEK and –N groups of the imidazole.

IEC plays a crucial role in deciding the water uptake and retention of water molecules inside the membranes. This in turn facilitates the water assisted proton conduction, however an excessive uptake would result in swelling with progressive decrease in mechanical strength of the membrane affecting the long term durability

DOI: 10.1201/9781003733744-7

of the PEM. Hence, a judicial balance between the two would lead to efficient and durable PEM. Therefore, an optimized membrane with minimum inbuilt chemical and mechanical stress forms a stable membrane electrode assembly (MEA) and efficiently contributes to achieve durable fuel cell performance in real situation.

7.1.2 Water and Methanol Uptake

For easy conduction of protons water inside the membrane plays a major role. In the case of functionalized PEMs, water acts as vehicle and provides the path for movement of protons as discussed in Chapter 2. In the case of DMFCs where methanol acts as a fuel, the uptake of methanol as liquid is also a matter of concern. Hence, for the evaluation of membrane, determination of liquid uptake is important. The liquid up take (water and methanol) capacity of the membranes is determined from Eqn. (7.2). Dried membranes are dipped in certain amount of water or methanol for sufficient time period at room temperature and the change in weight is measured with time until the weight becomes constant. However, many PEMs suffer from the problem of excessive swelling at high degree of functionalization leading to dimensionally unstable membranes. The amount of liquid intake is calculated from the difference in weight of dry and wet membranes.

$$\text{Liquid up take} = \left(W_{\text{wet}} - W_{\text{dry}}\right) \times 100 \,/\, W_{\text{dry}} \tag{7.2}$$

where, W_{wet} and W_{dry} represent the weight of film in gram under wet and dry conditions.

7.1.3 Thermal Properties

7.1.3.1 Thermal stability and phase transitions

Thermal stability of the membranes is very essential for the successful long-term operation of the fuel cell devices. The thermal stability of the membranes is determined by using thermo gravimetric analyzer (TGA). In TGA, the dry membranes are heated at a specified heating rate of 5°C–10°C in an inert atmosphere preferably in nitrogen atmosphere. The weight change with temperature is recorded. Thermal transitions of the membranes are determined by differential scanning calorimeter (DSC) where the melting point, glass transition temperature, and crystallinity of the polymer membranes are measured. Determination of thermal transitions has a great significance in the case of blend and composite membranes particularly to understand the different phase transitions occurring in the blend and composite membranes.

The thermo-mechanical stability of aromatic hydrocarbon based ionomers is higher than PFSA ionomers which makes them stable at higher temperature [3, 4]. Aromatic ionomers have glass transition temperature (T_g) in the range of 180°C–500°C which is higher compared to Nafion (T_g of around 130°C– in detail 160°C) [5–7]. This makes the membrane more stable during fuel cell operation at high temperature of 100°C–250°C. Aromatic hydrocarbon based ionomers such as poly(arylene pyridine) s (PAP) and polybenzimidazole (PBI) have glass transition temperature of more than 400°C and thus, can be used in high temperature conditions. At high temperature operating conditions, due to non-availability of proton conduction by water, fuel cells use phosphoric acid doped membranes due to the high boiling point of phosphoric acid (~212°C) as well as good proton transfer properties [8]. The only drawback is that phosphoric acid leaches out from ionomer over time and thus reducing the performance of cell and eventually leading to the cell failure [9, 10]. This limitation can be addressed by using functionalization of aromatic ionomers with phosphonic groups, though their proton conduction is highly dependent on the temperature [11]. Running fuel cells at ~150°C offers more benefits when compared to low temperature operating PEMFCs. These benefits include higher reaction rates, no water management problems, no humidification, high gas bulk diffusivity, reduced radiator size, and less sensitivity to impurities like CO. This makes high-temperature operating PEMFCs potentially cheaper and more efficient than low-temperature ones [3].

Imaan et al. investigated the thermal stability and mechanical properties of α-alumina/PVA composite PEM [12]. Figure 7.1(a) shows different stages of weight loss process in TGA thermo gram. In the first step, the membrane shows about 5% weight loss due to removal of remnant water from the membranes. After that, in the temperature range of around 260°C–320°C, it displays weight loss due to the side chain degradation of hydroxyl groups of PVA and finally over 320°C and up to 570°C, the main chain degradation occurs. For composite membranes containing 10%–30% α-alumina, the membranes are stable up to 260°C with good water retention ability as observed in the TGA. Figure 7.1(b) displays the effect of α-alumina in PVA matrix. The tensile strength of the membranes is observed to show huge increase with the enhancement in filler content.

Lochner et al. investigated the effect of temperature on the stability of commercial PEM [13]. During the operation of fuel cell vehicles, the device faces huge change in temperature from cold start to operation condition. The device reaches to temperature as low as −20°C during the cold start and during operation the temperature reaches to about +120°C depending of the type of fuel cell devices i.e., low, medium and high temperature fuel cells. Hence, it is very important to understand the effect of temperature on the thermal stability of the membranes. The thermal management procedure, challenges, and assessment of PEMFC devices are described in Chapter 8.

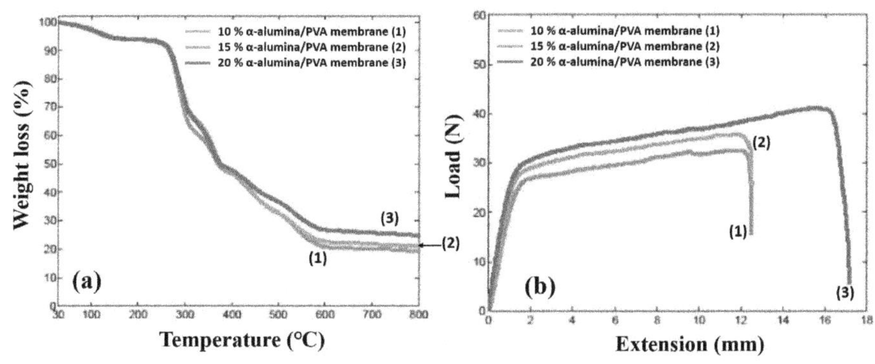

FIGURE 7.1 (a) Thermal stability and (b) mechanical properties of α-alumina/PVA composite PEM [12] (Reprinted with permission)

7.1.4 Mechanical properties

The Nafion is the most popular commercial polymer electrolyte membrane of DuPont consisting perfluorosulfonic acid (PFSA) moiety. It is highly utilized in fuel cells due to its high proton conductivity. Initially, the Nafion membranes utilized in the fuel cells were of high thickness ranging from 50–200 nm, thus, the mechanical stability of Nafion membrane was not a problem. Nowadays, thinner membranes of Nafion are used which leads to early failure of membrane. The membrane suffers from chemical and mechanical stresses due to the attack of radical or reactive species, swelling and dehydration of membrane. During the operation of fuel cell, hydration and dehydration cycle causes generation of holes and cracks within the membrane [14–16] resulting in the performance loss and cell failure. The tension is caused by the swelling and shrinkage of membrane in the presence of water and heat however, the membrane remains constrained due to the cell geometry and cannot freely expand [14, 15]. The major stress concentrates at the meeting point of membrane and MEA frame, specifically near the hydrogen entry point due to some hydrogen leakage across the membrane [17]. This leaked hydrogen reacts with oxygen to form thermal hotspots. The edges of membrane suffer from high stress and undergo uneven bending and stretching over the time leading to non-uniform deformations in the membrane. This leads to generation of holes and further propagation of the cracks during shut off and start-up of the fuel cell. High temperature and water make the condition worsen as the membrane becomes softer and hence it starts to swell more causing faster crack propagation and thinning of the membrane edges. All of this increases the mechanical stress inside the membrane eventually causing the membrane failure [18].

The mechanical properties of membranes are assessed by measuring tensile strength, modulus, elongation at break, tear resistance, bulging test, etc. Additionally, Dynamic Mechanical Analysis (DMA) technique is also used to measure thermomechanical transitions within the membrane. A dynamic mechanical study is conducted on Nafion membrane and loss tangent and tensile modulus are measured. The study revealed that, first transition at around ~110°C (alpha transition) is responsible for starting of free chain movement due to weak electrostatic network. A second transition at ~20°C (beta peak) is observed due to segmental motion of polymeric chains within the polymeric framework [19]. The study of rheological behavior of Nafion by DMA revealed that stress-strain curve is highly non-linear in the case of cycling loading [14, 15, 20]. This behavior can be effectively modelled as an intermolecular resistance parallel to elastic network resistance [14]. The intermolecular resistance gets weaker at high temperature and RH. The elastic network resistance which is responsible for the reorientation of polymer chains and it does not change much with temperature and RH. In reality, the evaluation is conducted by measuring the Young's modulus (E), Tensile strength (TS) and Break strain (BS) from the stress-strain curve obtained in the DMA and Universal Testing Machine (UTM) under controlled temperature and RH. The study revealed that both E and TS decrease with heat and RH, whereas, BS shows opposite behavior and increases with the decrease in heat and RH [21, 22]. The absorption of water results in swelling of membrane in all three directions, thus, the quantification of the volume change is done by determining the swelling ratio (*SR)* using Eqn. 7.3

$$SR = \frac{\Delta V_w}{V_{dry}} = (1+SR_{ip,1})(1+SR_{ip,2})(1+SR_{ip}) - 1 = \tilde{\rho} WU \tag{7.3}$$

where, ΔV_w and V_{dry} represent the volume of absorbed water and volume of dry membrane. The total volumetric swelling as swelling components in three directions is represented by $SR_{ip,1}$ and $SR_{ip,2}$ (linear extension in material plane) and SR_{ip} (linear extension in thickness direction). $\tilde{\rho}$ indicates water uptake (WU) to volumetric swelling.

During operation of the fuel cell, higher water uptake results in increased proton conduction however, higher water uptake increases the membrane swelling causing the situation to become worsen. This situation can be overcome by designing membranes that swell mostly in thickness direction than the in plane direction [23]. This will lead to increased water uptake with less stress in the direction of plane of the cell. This can be achieved by multilayer reinforcing the membrane with PTFE which provides tailored crystallinity at the interfaces of reinforcements. Another such approach includes adding a second phase to the membrane material. This increases the density and crystallinity of crosslinking sites and increases the value of Young's modulus and tensile strength. Such reinforcing materials include MOFs, covalent organic frameworks (COFs) and nanoparticles [24–27]. Jahangiri et al. investigated the mechanical properties of PBI and phosphoric acid doped PBI membranes [28]. The membranes are doped for 24–96 hours to observe the effect of doping on the mechanical properties of the doped membranes. The study reveals that the doping with phosphoric acid leads to the formation of H-bonding leading to an enhancement in tensile strength in the doped membranes. The PA doped membranes exhibit less ductility as compared to un-doped membranes (Figure 7.2).

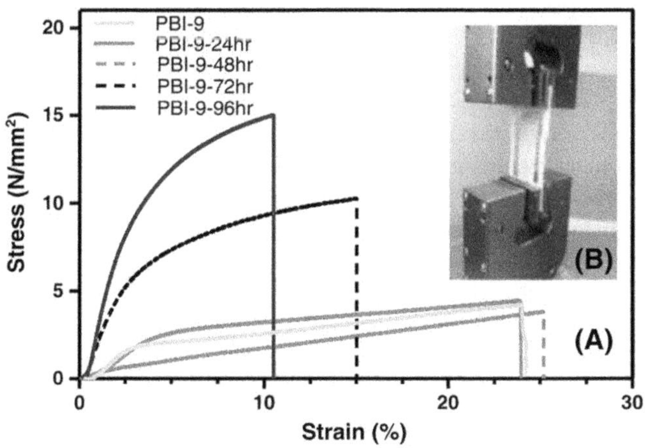

FIGURE 7.2 (a) Stress-strain plots of un-doped and phosphoric acid–doped electrospun membranes, (b) Set up of mechanical testing [28] (Reprinted with permission)

7.2 IMPEDANCE ANALYSIS

For a commercially viable energy storage product, fuel cell has to meet the standards set by US Department of Energy (DOE). According to the standards, a stationary fuel cell should be able to provide over 80% of power for a lifetime of 40,000 h including 8,000 h of uninterrupted power supply [29–31]. In order to achieve the targets, set by DOE, the PEM fuel cell need to be examined by suitable characterization techniques. Electrochemical analysis of a PEM fuel cell uses some specific techniques which include analysis of polarization curve, electrochemical impedance spectroscopy (EIS), cyclic voltammetry, and current interruption. EIS is able to distinguish the effect of different processes taking place simultaneously, effectively reveal the different mechanisms involved in the process such as, parallel reactions, adsorption, and multi-step reactions. As the mechanism of fuel cell is governed by different processes which may simultaneously overlap with each other, thus, impedance analysis is carried out to obtain comprehensive idea about such processes. EIS involves the measurement of impedance of a fuel cell over a given frequency range by applying an alternate voltage/current of a small amplitude. The impedance, phase, and magnitude of the obtained response at various frequencies reveal the kinetic and underlying electrochemical behavior of the fuel cell. By analysis of the various parameters obtained one can find out about the microscopic information, appropriate equivalent circuit, electrochemical parameters, contributions of different components to the fuel cell performance and determination of adsorption, charge transfer and mass transfer processes in both the catalyst and backing layer [32].

The EIS measurement can be carried out either in a three-electrode configuration or in a two-electrode arrangement. The three-electrode configuration offers material characterization and thus, helps in selection of material and component integration. In the three-electrode, the measurements are carried out using a working electrode, counter electrode and reference electrode in acidic [33, 34] or alkaline [35] electrolyte medium. The common reference electrodes include SCE, NHE, Ag/AgCl, or Hg/HgO. The selection of reference electrodes depends on the electrolyte medium in which the measurements are being taken. The working electrode is made up of a thin layer of catalyst casted on glassy carbon or pyrolytic graphite [36, 37].

However, the two-electrode configuration provides practical data on PEMFCs properties. In, a two-electrode configuration, the anode works as a combined assembly of reference and counter electrode. The hydrogen is passed through the anode and air is passed through the cathode. The only limitation with two-electrode configuration is that the measurement of single electrodes cannot be obtained or evaluated. But during operation, the polarization of anode is almost negligible in comparison with polarization of cathode. This diminishes the small limitation in the case of hydrogen as fuel [38].

The information of different components can also be obtained by an equivalent circuit fitting with the impedance spectra via employing different modes of measurement (potentiostatic mode or galvanostatic mode), at variable gas-feeding modes (H_2/H_2, O_2/O_2, H_2/O_2, H_2/air, H_2/N_2, etc.), or at varying working conditions e.g. humidity, temperature, and catalyst.

The impedance measurement of the system is carried out by a frequency response analyzer (FRA) connected to a DC load bank. FRA generates a sinusoidal wave which passes through the fuel cells and is superimposed on DC current. The generated signals are measured by FRA to produce the impedance spectra. There are two different modes for AC impedance measurement: (1) potentiostatic mode (2) galvanostatic mode. Both modes have their own advantages and their suitability depends upon the application. In potentiostatic mode, to obtain AC impedance of the cell, AC potential is applied while the response current is measured. A small variation in potential may result in large variation in current which causes overloading of the cell. Thus, it is essential to keep caution ensuring that the current does not exceed the given limits of potentiostat. Therefore, generally a small amplitude of 5–15 mV for AC signal is applied [39–41] over a frequency range of 1 mHz to 100 kHz [42–44]. The impedance can either be collected near open circuit voltage or at a DC load. Near open circuit voltage, the kinetics of electrode is dominant, Ohmic contribution is minimum, and mass transfer also takes place up to a minimum amount. At a DC load, the obtained impedance data represents the complete processes taking place at the electrode [45]. However, some measurements can only be conducted at OCV such as the impedance of a gas feeding cell. At low overpotential, interfacial charge transfer resistance is the main cause of resulting impedance. This interfacial charge transfer resistance is basically dependent on the potential of electrode. The relationship between electrode potential and interfacial charge transfer resistance is established by Tafel equation as given below [46]:

$$V \sim \log R_P^{-1} \tag{7.4}$$

As the overpotential increases, the impedance is mainly caused by movement of fuel, oxygen, proton, and water within the cell. The further increase in overpotential results in dominance of oxygen diffusion and gas diffusion in the case of air as an oxidant [47]. This behavior can be seen in the impedance plot, where the interfacial charge transfer resistance arc decreases with the increase in overpotential. This reflects the driving force in process of oxygen reduction [39].

In galvanostatic mode, to obtain AC impedance of the cell, AC current is applied while the response potential of the cell is measured. In contrast to potetiostatic mode, galvanostatic mode provides the control over the value of current passing through the cell. Thus, prevent the accidental overloading of potentiostat. The galvanostatic mode is advantageous for high-energy devices measurement to control high current levels. Additionally, during the impedance study of CO contamination galvanostatic mode is preferred as it gives the impedance of anode. In potentiostatic mode the impedance of anode and cathode varies with respect to the time, hence, it becomes difficult to accurately analyze the impedance of CO poisoning [48].

Apart from the various impedance of the cell that includes charge transfer and mass transfer resistance the impedance curve can reveal the proton transport characteristics through the proton exchange membrane. The Ohmic resistance is associated with the resistance faced by proton transfer through the proton exchange membrane. This is mainly governed by the proton conductivity of the electrolyte membrane. The polymer electrolyte membrane with high proton conductivity offers efficient ion transport resulting in low Ohmic resistance and improved overall performance.

Generally, proton conductivity of the membrane is evaluated along the plane which is called as in-plane conductivity [49–51]. However, the conduction of proton happens to take through the plane which is perpendicular to the plane of membrane. Additionally, material may exhibit morphological anisotropy which leads to inaccurate data estimation by in-place conductivity evaluation. The Nafion membrane majorly possess anisotropic morphology as described in several studies mentioned in literature [52–55]. The anisotropic morphology depends on the casting and pretreatment methods of the membrane. Distinct anisotropic morphology is displayed by the novel materials for proton exchange membrane, for, e.g., sulfonated polyimides resulting in anisotropic proton conductivity [56]. Thus, it becomes essential to evaluate through-plane proton conductivity by the impedance study of fuel cells. The through-plane proton conductivity has been measured by different methodology and apparatus in different reports [52, 53, 55, 57–61]. Ma et al. studied through plane conductivity of commercial Nafion membrane in which the assembly consists of 5μL of 5% Nafion dispersion coated on Pt sheets and hot pressed after being sandwiched between two electrodes [55]. The subsequent arrangement is made by two polyester-coated wire electrodes sandwiched between three Nafion membranes followed by hot pressing of the assembly. Another assembly is prepared by placing Pt on both sides of Nafion membrane [59]. In another study, Ramya et al. assembled the cell by placing Nafion membrane between two carbon paper electrodes [60]. Blachot et al. used mercury electrodes to evaluate the proton conductivity of sulfonated polyimide membrane. The measurement setups to evaluate both in-plane and through plane proton conductivity of membranes vary greatly [62]. In through plane conductivity measurement, the cell is built by PTFE blocks and thin Pt sheets are attached to the blocks as electrodes as shown in Figure 7.3(b). Nylon screws are used to fix the PTFE blocks and electrical pins connect the Pt electrode to the impedance analyzer. Before taking measurements, the system is calibrated by recording the impedance data of shorted cell and an open cell. Later on, these readings are used to nullify the stray inductance of the system in Z plot software. The proton exchange membrane is cut to a square shape of 5 mm × 5 mm and sandwiched between the Pt electrodes. To ensure the proper contact within membranes the cell is secured with screws. The compression pressure is monitored using Pressurex® which is a pressure indicating film works in the range of 350–1400 psi. The whole setup is plunged into deionized water and EIS data is recorded to evaluate the through plane proton conductivity of the membrane. For in-plane proton conductivity measurement, the cell is assembled using Pt sheets of 10 mm × 10 mm placed 5 mm apart and attached to PTFE block as show in Figure 7.3(a). The membrane of 10 mm × 10 mm is placed between the electrodes. The whole setup is submerged in water to maintain the hydration of membrane and the impedance data is recorded. The measurements are taken in two in-plane directions, i.e., X and Y and that conductivity (σ) is calculated according to Eqn. 7.5

$$\sigma = \frac{L}{R.A} \tag{7.5}$$

where, L, R and A represent the distance between electrodes, bulk resistance evaluated from impedance plot and cross-sectional area of membrane. In through-plane

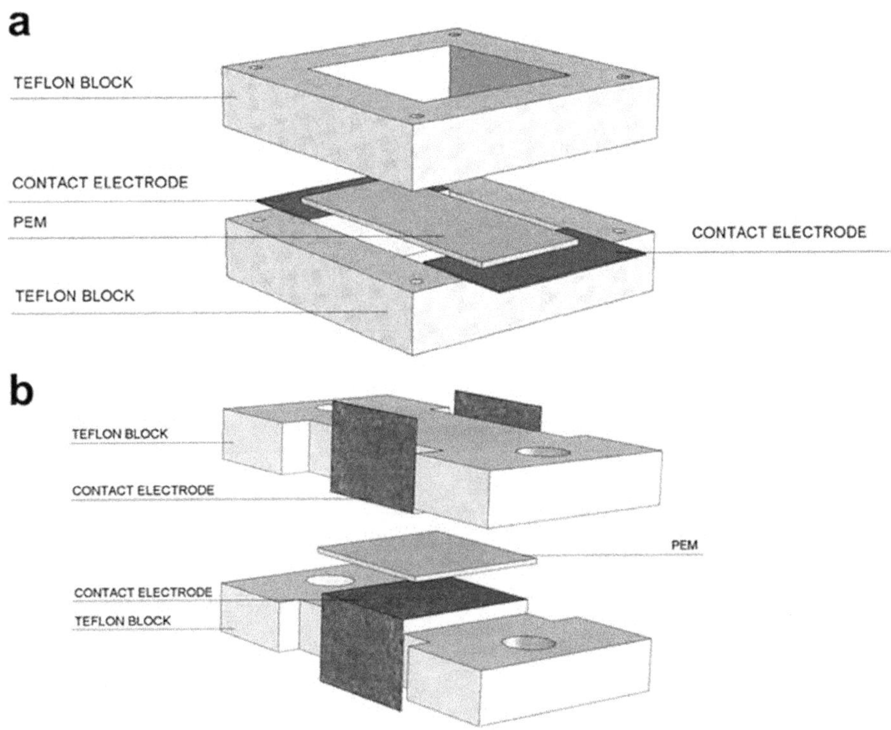

FIGURE 7.3 The measurement setups of (a) in-plane and (b) through-plane impedance cell [62] (Reprinted with permission)

measurement setup, the separation distance between electrodes is same as thickness of the electrolyte membrane.

7.3 FABRICATION OF MEA

The electrode membrane assembly (MEA) is an integrated system which consists of seven consecutive layers used to finally fabricate the functional PEM fuel cell. These seven layers of MEA is sequentially comprised of: (1) sealing gasket, (2) gas diffusion layer, (3) anode catalyst layer, (4) proton exchange membrane, (5) cathode catalyst layer, (6) gas diffusion layer, and (7) sealing gasket as shown in Figure 7.4. Each layer serves a specific function and alteration in any of these layers significantly affects the performance of the whole assembly. Proton exchange membrane also known as electrolyte membrane in the fuel cell, being the core component of MEA assembly facilitates the transfer of protons from cathode to anode. The membrane developed by DuPont is known by the trade name of Nafion and is made up of sulfonic acid containing perfluorinatedvinyl polyether side chain attached on polytetrafluoroethylene (PTFE) backbone [63].

Once the hydration of membrane takes place, it allows the transfer of protons. But the proton exchange membrane usually exists as a composite structure consisting of conducting polymer (Nafion) and anode or cathode catalyst layer (metal catalyst

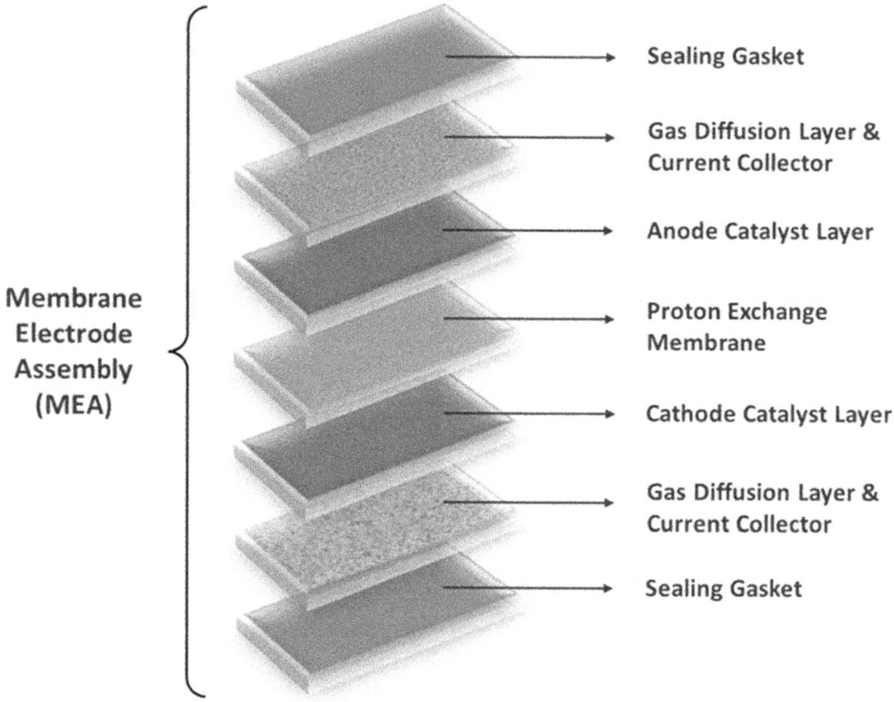

FIGURE 7.4 A seven layer MEA (Adapted from [66])

supported on carbon framework). The thickness of catalyst layer generally varies from 10 to 20 μm. This variation of thickness is based on the amount of catalyst loaded. So far, platinum is proven to be the best choice as a catalyst for both the electrodes. The selection and loading of catalyst is greatly influenced by the source of fuel. In the case of pure hydrogen fuel, the oxidation of hydrogen gas is a facile reaction which results in a small overpotential. Thus, the requirement of Pt catalyst is also small and loading is less. However, if reformate or methanol is used as a fuel source, pure Pt catalyst will suffer from CO poisoning. To minimize the effect of CO contamination on catalyst Pt-based alloys are used in place of pure Pt catalyst such as PtNi, PtRh, and PtRu alloys [64].

The optimum value of catalyst loading at cathode is 0.1–0.2 mg/cm^2. Catalyst particles in size range of approximately 5–20 nm are generally dispersed on the surface of carbon frame (Vulcan XC72 Cabot) for support. The use of nanoparticles provides with a relatively large specific surface area which leads to efficient turnover rate for per mg of catalyst. The catalyst layer is generally fabricated either by electrodeposition method or directly imprinted on proton exchange membrane. In the electrodeposition method, the catalyst is first electrodeposited over a conducting and porous carbon material and then hot pressed on both side of polymer membrane at 132°C for 2–3 minutes. In the next method, platinum containing carbon catalyst is directly applied on polymer membrane by various steps such as rolling, spraying, or screen printing [65]. These processing steps may affect the efficacy of proton exchange membrane, catalyst layer, and their interface. Thus, these steps should be performed carefully to avoid variation in materials leading to a stable performance. The catalyst-electrode composite layer not only works as an active site for charge transfers but also provide the conduction path for protons and electrons. The protons move through the PEM and electrons via carbon to external circuit.

The gas diffusion layer forms a protective covering over the anode and cathode catalyst layers. The function of gas diffusion layer is to uniformly distribute the reactant molecules from the fuel source on the surface of catalyst layer. Additionally, the transportation of generated water and electrons are also ensured by gas diffusion layer. The dissipation of heat of reaction is also conducted through gas diffusion layer. The material used for gas diffusion layer include carbon paper (Toray paper) and PTFE as it provides hydrophobicity to the layer. The commercial one possesses in plane resistivity of 25–100 mΩ cm and through plane resistivity of 6–20 mΩ cm. The microstructure of the layer affects the value of these resistances which in turn dictates uniformity of reaction and cell resistance. The last layers on both sides comprises of sealing and gaskets which prevents the leaking and over-compression. The sealing thickness determines the compression over gas diffusion layer and thus, provide with a compact design.

7.4 CONSTRUCTION OF EQUIVALENT CIRCUIT

The EIS spectra provide the information about the on-going processes that take place during the proton transfer and quantification can be done by applying a fitting equivalent circuit diagram. Thus, to represent the overall impedance in form of circuit diagram the understanding of different components contributing to the impedance of cell becomes essential. In the case of through-plane conductivity measurement, assembly of cell involves an electrode/electrolyte interface. This interface is responsible for the formation of double layer, thereby contributing in double layer capacitance and a resistance to the flow of charges. Thus, this combination of capacitance and resistance is displayed by their parallel combination in circuit. In Figure 7.5(a), C_{int1} and R_{int1} represent the capacitance and resistance responsible for

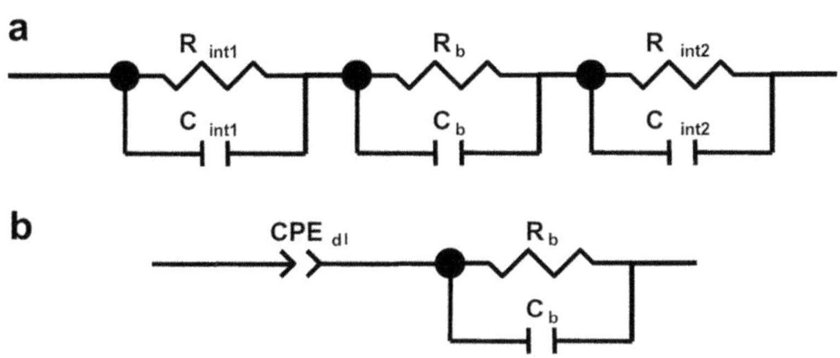

FIGURE 7.5 Equivalent circuits for the PEM in through-plane analysis: (a) Idealized circuit, (b) Simplified equivalent circuit for fitting of the experimental data. [62] (Reprinted with permission)

first electrode/membrane interface and C_{int2} and R_{int2} represent the capacitance and resistance responsible for second electrode/membrane interface [62]. The polymer membrane itself possesses the bulk membrane capacitance, C_b, and a bulk resistance that opposes the flow of proton, R_b, across the membrane. As faradic charges are not involved in the process, the electrode/membrane interface transfer is purely capacitive in nature, thus, resistance can be removed from the circuit diagram [67, 68]. Additionally, pure capacitive behavior results in the line parallel to Z″ axis in the Nyquist plot which is ideally not found in the solid electrode surface due to irregularity of the electrode surface. The capacitance deviates from its ideal value depending upon phase-angle frequency. Hence, this behavior is represented by a constant phase element (CPE) as shown in Figure 7.5(b).

The interfacial impedance can be evaluated by using Eqn. 7.6

$$Z(\omega) = \frac{1}{(i\omega C_{dl})^n} \tag{7.6}$$

Here, the value of ω does not affects the value of C_{dl} and the value of n lies between 0.5 to 1 ($n = 1$ represent ideal capacitor) [67]. At high frequency, the interfacial resistance caused by proton transfer is relatively very small as the distance travelled by protons is in Å (thickness of Helmholtz layer). Hence, the final circuit diagram of through plane cell assembly as combination of interfacial charge transfer and membrane transfer is expressed in Figure 7.5(b) and the expression for impedance is given in Eqn. 7.7

$$Z(\omega) = \left(\frac{1}{R_b} + i\omega C_b\right)^{-1} - \frac{1}{(i\omega C_{dl})^n} \tag{7.7}$$

The EIS data recorded is expected to give a semi-circle arc in high frequency region followed by a line inclined toward lower frequency zone corresponding to bulk membrane resistance and interfacial capacitance [30]. The value of bulk membrane resistance can simply be obtained from the diameter of the semi-circle, but in most of the cases a perfect complete semi-circle does not exist. In such cases, the estimation of bulk resistance can either be done by a liner extrapolation of impedance spike and intercept of Z′ axis is considered as the value of bulk resistance [69–72] or by equivalent circuit fitting. Figure 7.6 displays the Nyquist plot for commercial Nafion membrane fitted with impedance data obtained from the equivalent circuit using simulation. The Nyquist plot is obtained at frequency range varied from 100 Hz to 10 MHz, 100 Hz to 20 MHz, and 100 Hz to 100 MHz as observed in Figure 7.6. The semicircle passes through the origin of the coordinate and hence the diameter is considered as the resistance offered by the membrane. The simulated data points taken at various frequency ranges matches well with the experimental results as evidenced in the fitted plot.

In one such study, the data extracted from both the methods is compared for Nafion 117 [62]. The bulk resistance value obtained from linear extrapolation resulted in a value of 1.21 Ω whereas from equivalent circuit analysis the value is 1.18 Ω. Similarly, for Nafion 112, the linear extrapolation and circuit analysis resulted in

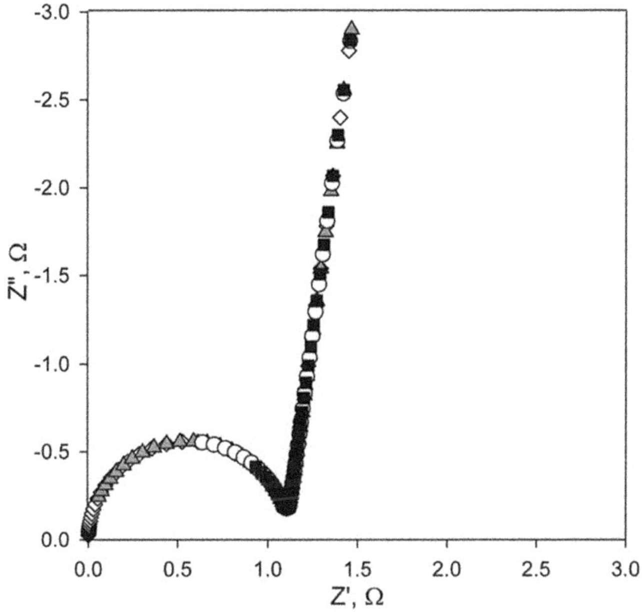

FIGURE 7.6 Nyquist plots fitted with impedance data obtained from the equivalent circuit using simulation; frequency range: '■' for 100 Hz to 10 MHz, 'O' for 100 Hz to 20 MHz, '▲' for 100 Hz to 100 MHz and '◊' for 100 Hz to 1 GHz [62] (Reprinted with permission)

bulk resistance value of 0.45 Ω and 0.50 Ω, respectively [62]. From both the methods, an overall deviation of 3% and 10% in bulk resistance value are obtained for Nafion 117 and Nafion 112, respectively.

7.5 PERFORMANCE EVALUATION OF DEVICE

The electrochemical performance evaluation of PEMFCs includes the study on the behavior of cell at given conditions to assess the key parameters involved to know the suitability of cell for practical application. The common tests for evaluation of PEMFCs performance include Polarization curve, Cyclic voltammetry (CV), Electrochemical Impedance Spectroscopy (EIS). The tests are performed at a specific cell pressure, cell temperature, and gases temperature. The flow rate of the fuel fed to the cell can either be fixed or varied according to the cell current which is known as load based flow control.

Polarization Curve: The polarization data is obtained by measuring the current output at varying voltage. In the curve, voltage is plotted against the current density. The resistance of the cell can be measured by current interrupt method. The polarization curve for a typical PEMFC devices is displayed in Figure 7.7 [73]. The curve indicates a considerable variation from ideal potential of 1.23 V. The losses responsible for this deviation are due to the electrode kinetics, reaction overpotential, etc. The slow oxygen reduction reaction (ORR) has also a remarkable effect. The ORR is found to be about six-order magnitude lower as compared to the hydrogen oxidation

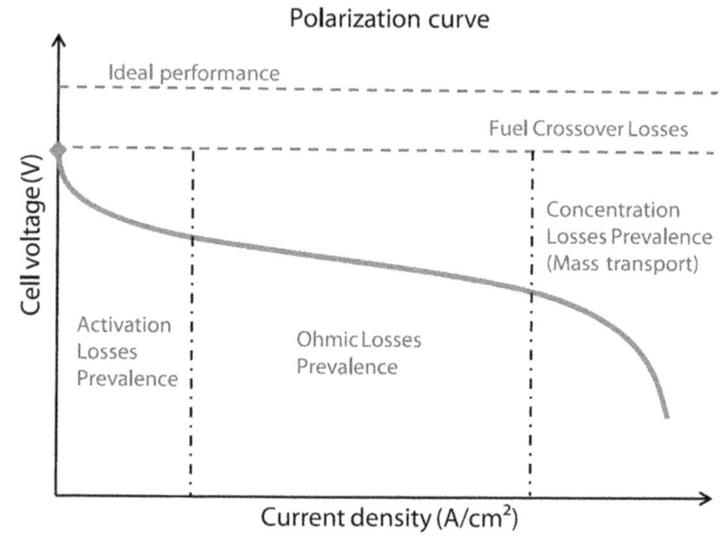

FIGURE 7.7 Polarization curve of PEMFC [73] *(Reprinted with permission)*

reaction limiting the efficient cell performance in the PEMFC devices. To cope with this problem, platinum catalyst is required. However, the huge cost of Pt and low durability demand development of alternative catalyst using non-precious metals. The activation polarization is observed due to exhausted reaction kinetics across the electrode. The Ohmic losses are evident due to resistance offered during the conduction of ions in the electrolyte. The Ohmic losses can be handled either by keeping lower electrode separation and/or by improving the proton conductivity of the PEM. During the operation of the fuel cell, as the electrochemical reaction proceeds, the reactants are consumed. This results in a loss of potential owing to the drop in the initial concentration, and a gradient is observed in the concentration profile. The concentration polarization is observed due to various factors such as diffusion and dissolution of the reactants and products and diffusion in the electrode pores.

Cyclic Voltammetry: In cyclic voltammetry, the anode serves as combination of reference and counter electrode. Sweeping of the cell is carried out over a potential range at a specific potential scan rate in both forward and reverse directions. The current output is recorded and plotted against the applied potential. The CV curve provides an idea about the total charge adsorbed. This includes the determination of hydrogen oxidation reaction and the final voltage. The electrochemical surface area (ECSA in m^2/g) of the catalyst is evaluated by using hydrogen adsorption charge (Q_{Hads} in C) which is associated with adsorbed hydrogen over the catalyst surface (Pt: 210 $\mu C/cm^2$) as given in Eqn. 7.8.

$$\text{ESCA} = \frac{Q_{Hads}}{210\ L_{cath} A_{geo}} \tag{7.8}$$

Here, L_{cath} represents loading of cathode, and A_{geo} represents the geometrical area of electrode.

Electrochemical Impedance Spectroscopy (EIS): The impedance measurement of the system is carried out by a frequency response analyzer (FRA) connected to a DC load bank. FRA generates a sinusoidal wave which passes through the fuel cells and is superimposed on DC current. The generated signals are measured by FRA to produce the impedance spectra. The data is plotted as imaginary impedance vs. real impedance (knows as Nyquist plot). This plot helps in evaluation of various resistance working simultaneously in the fuel cell. The acting resistance includes Ohmic, charge transfer, and mass transfer resistance.

7.6 CHAPTER SUMMARY

This chapter discusses the methods for evaluation of membrane properties by various methods such as ion exchange capacity, liquid uptake properties, and thermal and mechanical stability. The proton conductivity is a vital parameter to estimate the overall performance of the device. The proton conductivity is measured using AC impedance analysis where the conductivity is measured at different frequency levels and presented as a function of real and imaginary impedance. The experimental data is required to be fitted with the construction of the equivalent circuit depending on the interfaces of a specific membrane system to find out proton conductivity of the membranes in S cm^{-1}. The proton conductivity is evaluated both at room and higher temperature to see the effect of temperature on the proton conduction ability. Finally, after the fabrication of the membrane electrode assembly, the device is subjected to performance evaluation by electrochemical analysis.

REFERENCES

[1] Meng H, Song J, Guan P, Wang H, Zhao W, Zou Y, et al. High ion exchange capacity perfluorosulfonic acid resine proton exchange membrane for high temperature applications in polymer electrolyte fuel cells. *J Power Sources* 2024;602:234205. https://doi.org/10.1016/j.jpowsour.2024.234205.

[2] Tellez-Cruz MM, Escorihuela J, Solorza-Feria O, Compañ V. Proton exchange membrane fuel cells (PEMFCs): Advances and challenges. *Polymers (Basel)* 2021;13:3064–117. https://doi.org/10.3390/polym13183064.

[3] Rosli RE, Sulong AB, Daud WRW, Zulkifley MA, Husaini T, Rosli MI, et al. A review of high-temperature proton exchange membrane fuel cell (HT-PEMFC) system. *Int J Hydrogen Energy* 2017;42:9293–314. https://doi.org/10.1016/j.ijhydene.2016.06.211.

[4] Qu E, Hao X, Xiao M, Han D, Huang S, Huang Z, et al. Proton exchange membranes for high temperature proton exchange membrane fuel cells: Challenges and perspectives. *J Power Sources* 2022;533:231386. https://doi.org/10.1016/j.jpowsour.2022.231386.

[5] Roy A, Hickner MA, Lane O, McGrath JE. Investigation of membrane electrode assembly (MEA) processing parameters on performance for wholly aromatic hydrocarbon-based proton exchange membranes. *J Power Sources* 2009;191:550–4. https://doi.org/10.1016/j.jpowsour.2009.02.059.

[6] Borjigin H, Stevens KA, Liu R, Moon JD, Shaver AT, Swinnea S, et al. Synthesis and characterization of polybenzimidazoles derived from tetraaminodiphenylsulfone for high temperature gas separation membranes. *Polymer (Guildf)* 2015;71:135–42. https://doi.org/10.1016/j.polymer.2015.06.021.

[7] Higashihara T, Matsumoto K, Ueda M. Sulfonated aromatic hydrocarbon polymers as proton exchange membranes for fuel cells. *Polymer (Guildf)* 2009;50:5341–57. https://doi.org/10.1016/j.polymer.2009.09.001.
[8] Vilčiauskas L, Tuckerman ME, Bester G, Paddison SJ, Kreuer K-D. The mechanism of proton conduction in phosphoric acid. *Nat Chem* 2012;4:461–6. https://doi.org/10.1038/nchem.1329.
[9] Lee AS, Choe Y-K, Matanovic I, Kim YS. The energetics of phosphoric acid interactions reveals a new acid loss mechanism. *J Mater Chem A Mater* 2019;7:9867–76. https://doi.org/10.1039/C9TA01756A.
[10] Lim KH, Lee AS, Atanasov V, Kerres J, Park EJ, Adhikari S, et al. Protonated phosphonic acid electrodes for high power heavy-duty vehicle fuel cells. *Nat Energy* 2022;7:248–59. https://doi.org/10.1038/s41560-021-00971-x.
[11] Atanasov V, Lee AS, Park EJ, Maurya S, Baca ED, Fujimoto C, et al. Synergistically integrated phosphonated poly(pentafluorostyrene) for fuel cells. *Nat Mater* 2021;20:370–7. https://doi.org/10.1038/s41563-020-00841-z.
[12] Imaan D ul, Mir FQ, Ahmad B. Proton exchange membrane based on poly (vinyl alcohol) as support and alpha alumina as filler and its performance in direct methanol fuel cell. *J Environ Chem Eng* 2021;9:106119–27. https://doi.org/10.1016/j.jece.2021.106119.
[13] Lochner T, Kluge RM, Fichtner J, El-Sayed HA, Garlyyev B, Bandarenka AS. Temperature effects in polymer electrolyte membrane fuel cells. *Chem Electro Chem* 2020;7:3545–68. https://doi.org/10.1002/celc.202000588.
[14] Silberstein MN, Boyce MC. Hygro-thermal mechanical behavior of Nafion during constrained swelling. *J Power Sources* 2011;196:3452–60. https://doi.org/10.1016/j.jpowsour.2010.11.116.
[15] Kusoglu A, Karlsson AM, Santare MH, Cleghorn S, Johnson WB. Mechanical response of fuel cell membranes subjected to a hygro-thermal cycle. *J Power Sources* 2006;161:987–96. https://doi.org/10.1016/j.jpowsour.2006.05.020.
[16] Dafalla AM, Jiang F. Stresses and their impacts on proton exchange membrane fuel cells: A review. *Int J Hydrogen Energy* 2018;43:2327–48. https://doi.org/10.1016/j.ijhydene.2017.12.033.
[17] Xing Y, Li H, Avgouropoulos G. Research progress of proton exchange membrane failure and mitigation strategies. *Materials* 2021;14:2591–607. https://doi.org/10.3390/ma14102591.
[18] Kusoglu A, Santare MH, Karlsson AM. Aspects of fatigue failure mechanisms in polymer fuel cell membranes. *J Polym Sci B Polym Phys* 2011;49:1506–17. https://doi.org/10.1002/polb.22336.
[19] Yeo SC, Eisenberg A. Physical properties and supermolecular structure of perfluorinated ion-containing (nafion) polymers. *J Appl Polym Sci* 1977;21:875–98. https://doi.org/10.1002/app.1977.070210401.
[20] Solasi R, Zou Y, Huang X, Reifsnider K. A time and hydration dependent viscoplastic model for polyelectrolyte membranes in fuel cells. *Mech Time Depend Mater* 2008;12:15–30. https://doi.org/10.1007/s11043-007-9040-7.
[21] Lu Z, Lugo M, Santare MH, Karlsson AM, Busby FC, Walsh P. An experimental investigation of strain rate, temperature and humidity effects on the mechanical behavior of a perfluorosulfonic acid membrane. *J Power Sources* 2012;214:130–6. https://doi.org/10.1016/j.jpowsour.2012.04.094.
[22] Tang Y, Karlsson AM, Santare MH, Gilbert M, Cleghorn S, Johnson WB. An experimental investigation of humidity and temperature effects on the mechanical properties of perfluorosulfonic acid membrane. *Mater Sci Eng: A* 2006;425:297–304. https://doi.org/10.1016/j.msea.2006.03.055.
[23] Kusoglu A, Weber AZ. New insights into perfluorinated sulfonic-acid ionomers. *Chem Rev* 2017;117:987–1104. https://doi.org/10.1021/acs.chemrev.6b00159.

[24] Çalı A, Şahin A, Ar İ. Incorporating sepiolite and kaolinite to improve the performance of SPEEK composite membranes for proton exchange membrane fuel cells. *Can J Chem Eng* 2020;98:892–904. https://doi.org/10.1002/cjce.23681.
[25] Mohsenpour S, Kamgar A, Esmaeilzadeh F. Investigation the effect of TiO2 nanoparticles on proton exchange membrane of sPEEK used as a fuel cell electrolyte based on phase diagram. *J Inorg Organomet Polym Mater* 2018;28:63–72. https://doi.org/10.1007/s10904-017-0723-5.
[26] Rynkowska E, Fatyeyeva K, Marais S, Kujawa J, Kujawski W. Chemically and thermally crosslinked PVA-based membranes: Effect on swelling and transport behavior. *Polymers (Basel)* 2019;11:1799–816. https://doi.org/10.3390/polym11111799.
[27] Upadhyaya L, Chiao Y-H, Wickramasinghe SR, Qian X. Cu(I/ II) Metal–organic frameworks incorporated nanofiltration membranes for organic solvent separation. *Membranes (Basel)* 2020;10:313–29. https://doi.org/10.3390/membranes10110313.
[28] Jahangiri S, Aravi İ, Işıkel Şanlı L, Menceloğlu YZ, Özden-Yenigün E. Fabrication and optimization of proton conductive polybenzimidazole electrospun nanofiber membranes. *Polym Adv Technol* 2018;29:594–602. https://doi.org/10.1002/pat.4169.
[29] Healy J, Hayden C, Xie T, Olson K, Waldo R, Brundage M, et al. Aspects of the chemical degradation of PFSA ionomers used in PEM fuel cells. *Fuel Cells* 2005;5:302–8. https://doi.org/10.1002/fuce.200400050.
[30] Knights SD, Colbow KM, St-Pierre J, Wilkinson DP. Aging mechanisms and lifetime of PEFC and DMFC. *J Power Sources* 2004;127:127–34. https://doi.org/10.1016/j.jpowsour.2003.09.033.
[31] LaConti AB, Hamdan M, McDonald RC. *Handbook of fuel cells–fundamentals, technology and applications.* vol. 3. Wiley; 2003.
[32] Yuan X, Sun JC, Wang H, Zhang J. AC impedance diagnosis of a 500W PEM fuel cell stack. *J Power Sources* 2006;161:929–37. https://doi.org/10.1016/j.jpowsour.2006.07.020.
[33] Schulze M, Wagner N, Steinhilber G. *Characterization and basic research investigations at PEFC electrodes and MEA*, 1996. Fuel Cell Seminar 1996, Orlando, FL (USA). https://doi.org/10.2172/460318.
[34] Easton EB, Pickup PG. An electrochemical impedance spectroscopy study of fuel cell electrodes. *Electrochim Acta* 2005;50:2469–74. https://doi.org/10.1016/j.electacta.2004.10.074.
[35] Genies L, Bultel Y, Faure R, Durand R. Impedance study of the oxygen reduction reaction on platinum nanoparticles in alkaline media. *Electrochim Acta* 2003;48:3879–90. https://doi.org/10.1016/S0013-4686(03)00525-5.
[36] Antoine O, Bultel Y, Durand R. Oxygen reduction reaction kinetics and mechanism on platinum nanoparticles inside Nafion®. *J Electroanal Chem* 2001;499:85–94. https://doi.org/10.1016/S0022-0728(00)00492-7.
[37] Song C, Zhang L, Zhang J. Reversible one-electron electro-reduction of O2 to produce a stable superoxide catalyzed by adsorbed Co(II) hexadecafluoro-phthalocyanine in aqueous alkaline solution. *J Electroanal Chem* 2006;587:293–8. https://doi.org/10.1016/j.jelechem.2005.11.025.
[38] Springer TE, Zawodzinski TA, Wilson MS, Gottesfeld S. Characterization of polymer electrolyte fuel cells using AC impedance spectroscopy. *J Electrochem Soc* 1996;143:587–99. https://doi.org/10.1149/1.1836485.
[39] Romero-Castañón T, Arriaga LG, Cano-Castillo U. Impedance spectroscopy as a tool in the evaluation of MEA's. *J Power Sources* 2003;118:179–82. https://doi.org/10.1016/S0378-7753(03)00085-5.
[40] Silva RF, De Francesco M, Pozio A. Tangential and normal conductivities of Nafion® membranes used in polymer electrolyte fuel cells. *J Power Sources* 2004;134:18–26. https://doi.org/10.1016/j.jpowsour.2004.03.028.

[41] Abe T, Shima H, Watanabe K, Ito Y. Study of PEFCs by AC impedance, current interrupt, and dew point measurements. *J Electrochem Soc* 2004;151:A101–5. https://doi.org/10.1149/1.1630589.
[42] Wagner N. Characterization of membrane electrode assemblies in polymer electrolyte fuel cells using a.c. impedance spectroscopy. *J Appl Electrochem* 2002;32:859–63. https://doi.org/10.1023/A:1020551609230.
[43] Ciureanu M. Effects of Nafion® Dehydration in PEM fuel cells. *J Appl Electrochem* 2004;34:705–14. https://doi.org/10.1023/B:JACH.0000031102.32521.c6.
[44] Diard J-P, Glandut N, Le Gorrec B, Montella C. Impedance measurement of each cell of a 10 W PEMFC stack under load. *J Electrochem Soc* 2004;151:A2193–7. https://doi.org/10.1149/1.1815152.
[45] Ahn S, Bruce J. Tatarchuk. Composite electrode structures for fuel cell applications, Vol. 3. 1990. IECEC–90; *Proceedings of the 25th Intersociety Energy Conversion Engineering Conference*, New York, American Institute of Chemical Engineers; 1990, p. 287–292.
[46] Ihonen J, Jaouen F, Lindbergh G, Lundblad A, Sundholm G. Investigation of mass-transport limitations in the solid polymer fuel cell cathode. *J Electrochem Soc* 2002;149:A448–54. https://doi.org/10.1149/1.1456917.
[47] Li G, Pickup PG. Ionic conductivity of PEMFC electrodes. *J Electrochem Soc* 2003;150:C745–52. https://doi.org/10.1149/1.1611493.
[48] Wagner N, Gülzow E. Change of electrochemical impedance spectra (EIS) with time during CO-poisoning of the Pt-anode in a membrane fuel cell. *J Power Sources* 2004;127:341–7. https://doi.org/10.1016/j.jpowsour.2003.09.031.
[49] Zawodzinski TA, Neeman M, Sillerud LO, Gottesfeld S. Determination of water diffusion coefficients in perfluorosulfonate ionomeric membranes. *J Phys Chem* 1991;95:6040–4. https://doi.org/10.1021/j100168a060.
[50] Fontanella J, Mclin M, Wintersgill M, Calame J, Greenbaum S. Electrical impedance studies of acid form NAFION® membranes. *Solid State Ion* 1993;66:1–4. https://doi.org/10.1016/0167-2738(93)90020-4.
[51] Zawodzinski TA, Derouin C, Radzinski S, Sherman RJ, Smith VT, Springer TE, et al. Water uptake by and transport through Nafion® 117 membranes. *J Electrochem Soc* 1993;140:1041–7. https://doi.org/10.1149/1.2056194.
[52] Gardner CL, Anantaraman AV. Measurement of membrane conductivities using an open-ended coaxial probe. *J Electroanal Chem* 1995;395:67–73. https://doi.org/10.1016/0022-0728(95)04072-V.
[53] Gardner CL, Anantaraman AV. Studies on ion-exchange membranes. II. Measurement of the anisotropic conductance of Nafion®. *J Electroanal Chem* 1998;449:209–14. https://doi.org/10.1016/S0022-0728(97)00408-7.
[54] Mauritz KA, Moore RB. State of understanding of nafion. *Chem Rev* 2004;104:4535–86. https://doi.org/10.1021/cr0207123.
[55] Ma S, Siroma Z, Tanaka H. Anisotropic conductivity over in-Plane and thickness directions in nafion-117. *J Electrochem Soc* 2006;153:A2274–81. https://doi.org/10.1149/1.2357727.
[56] Blachot J. Anisotropy of structure and transport properties in sulfonated polyimide membranes. *J Memb Sci* 2003;214:31–42. https://doi.org/10.1016/S0376-7388(02)00522-7.
[57] Elabd YA, Walker CW, Beyer FL. Triblock copolymer ionomer membranes. *J Memb Sci* 2004;231:181–8. https://doi.org/10.1016/j.memsci.2003.11.019.
[58] Wintersgill MC, Fontanella JJ. Complex impedance measurements on Nafion. *Electrochim Acta* 1998;43:1533–8. https://doi.org/10.1016/S0013-4686(97)10049-4.
[59] Parthasarathy A, Davé B, Srinivasan S, Appleby AJ, Martin CR. The platinum microelectrode/nafion interface: An electrochemical impedance spectroscopic analysis of oxygen reduction kinetics and nafion characteristics. *J Electrochem Soc* 1992;139:1634–41. https://doi.org/10.1149/1.2069469.

[60] Ramya K, Velayutham G, Subramaniam CK, Rajalakshmi N, Dhathathreyan KS. Effect of solvents on the characteristics of Nafion®/PTFE composite membranes for fuel cell applications. *J Power Sources* 2006;160:10–7. https://doi.org/10.1016/j.jpowsour.2005.12.082.
[61] Slade S, Campbell SA, Ralph TR, Walsh FC. Ionic conductivity of an extruded nafion 1100 EW series of membranes. *J Electrochem Soc* 2002;149:A1556–64. https://doi.org/10.1149/1.1517281.
[62] Soboleva T, Xie Z, Shi Z, Tsang E, Navessin T, Holdcroft S. Investigation of the through-plane impedance technique for evaluation of anisotropy of proton conducting polymer membranes. *J Electroanal Chem* 2008;622:145–52. https://doi.org/10.1016/j.jelechem.2008.05.017.
[63] DuPont Membrane Electrode Assemblies. Https://WwwDupontCom/Fuelcells/Products 2Html n.d.
[64] Ralph TR, Hogarth PH. Catalysis for low temperature fuel cells. *Platinum Met Rev* 2002;46:117–35.
[65] Wilson MS, Gottesfeld S. High performance catalyzed membranes of ultra-low pt loadings for polymer electrolyte fuel cells. *J Electrochem Soc* 1992;139:L28–30. https://doi.org/10.1149/1.2069277.
[66] Weidner JW, Sethuraman VA, Van Zee JW. Engineering a membrane electrode assembly. *Electrochem Soc Interface* 2003;12:40–3. https://doi.org/10.1149/2.F07034IF.
[67] Wang JC. Model for impedance of a solid ionic conductor sandwiched between blocking electrodes. *Electrochim Acta* 1993;38:2111–4. https://doi.org/10.1016/0013-4686(93)80348-4.
[68] Karthikeyan A, Vinatier P, Levasseur A. Study of lithium glassy solid electrolyte/electrode interface by impedance analysis. *Bull Mater Sci* 2000;23:179–83. https://doi.org/10.1007/BF02719906.
[69] Blachot J. Anisotropy of structure and transport properties in sulfonated polyimide membranes. *J Memb Sci* 2003;214:31–42. https://doi.org/10.1016/S0376-7388(02)00522-7.
[70] Wu GM, Lin SJ, Yang CC. Preparation and characterization of high ionic conducting alkaline non-woven membranes by sulfonation. *J Memb Sci* 2006;284:120–7. https://doi.org/10.1016/j.memsci.2006.07.025.
[71] Yang C-C, Lin S-J. Alkaline composite PEO–PVA–glass-fibre-mat polymer electrolyte for Zn–air battery. *J Power Sources* 2002;112:497–503. https://doi.org/10.1016/S0378-7753(02)00438-X.
[72] Wu GM, Lin SJ, Yang CC. Preparation and characterization of PVA/PAA membranes for solid polymer electrolytes. *J Memb Sci* 2006;275:127–33. https://doi.org/10.1016/j.memsci.2005.09.012.
[73] Jouin M, Gouriveau R, Hissel D, Péra M-C, Zerhouni N. Prognostics and health management of PEMFC – State of the art and remaining challenges. *Int J Hydrogen Energy* 2013;38:15307–17. https://doi.org/10.1016/j.ijhydene.2013.09.051.

8 Durability and associated challenges in PEMs

8.1 WHY DURABILITY ASSESSMENTS IN PEMFCS ARE IMPORTANT?

The adoption of green technology at large scale needs cost-effective and durable devices capable of competing with the existing technologies available in the market. The long-term durability of the PEMFCs is of utmost importance in order to achieve long life as well as to avoid the need of frequent replacement. The sustainability concepts suggest that a premature failure would not only make the process economically unfavorable but also lead to material wastage and energy consumption associated with the replacement resulting in the violation of the goal toward sustainable development. Thus, the durability of PEMFC is not just a technological demand but also extremely necessary toward achieving and contributing to the sustainability. In this regards, remarkable durability of PEMFCs is essential for their commercialization in stationary as well as transportation sectors. However, durability is the major issue for commercialization in stationary and transportation sectors of PEMFCs where the growth is maximum at present. There are various factors that contribute to overall durability of PEMFCs. In this chapter, we will be discussing all these factors briefly in the subsequent section. There are many test protocols available to assess the lifetime more precisely so that the durability issues can be resolved. To understand the life cycle of PEMFC, the decay in voltage and loss in power with operation cycles are assessed. The interpretation of durability data involves assessment of decay in voltage, loss in power with operation cycles and hours to understand the life cycle of PEMFC. These analyses give a solid foundation for testing of the various kinds of degradation occurred in components of PEMFCs during long operation conditions.

8.2 FACTORS AFFECTING THE DURABILITY OF PEM

Cost and durability remain two of the primary challenges in the advancement of fuel cell technology. These factors are closely interrelated, necessitating a well-optimized balance between them for practical application. Among various fuel cell types, the long-term stability of PEMFCs has been a critical concern since the inception of research in this field. The operational lifetime of a PEMFC is influenced by several degradation mechanisms, including membrane thinning, catalyst particle dissolution, and corrosion of the catalyst support material. Additionally, the equivalent weight of the proton exchange membrane plays a significant role in determining its performance and stability. While a reduction in equivalent weight can enhance

DOI: 10.1201/9781003733744-8

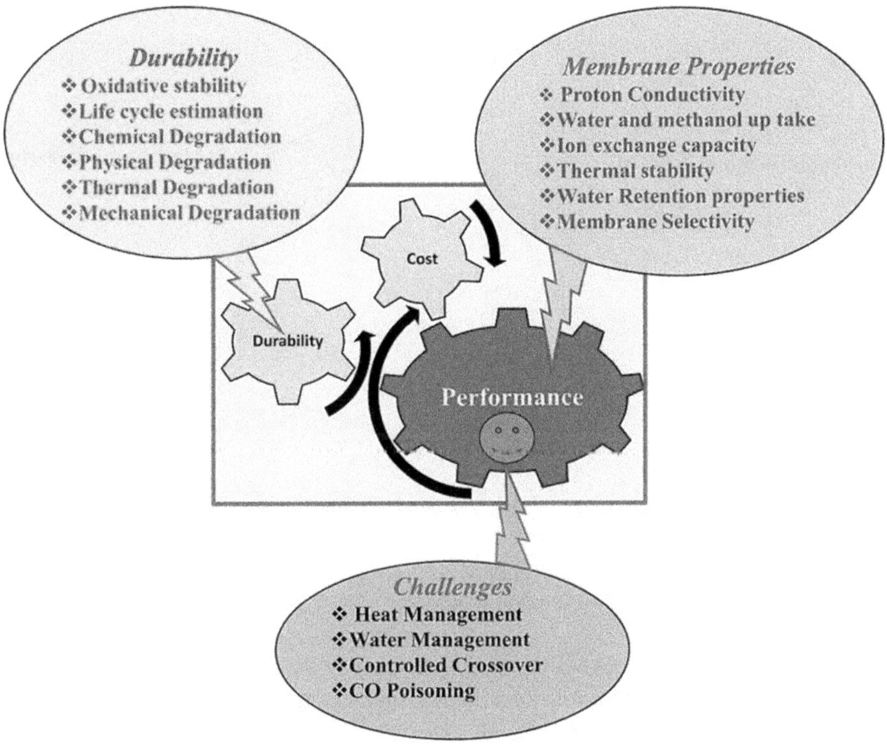

FIGURE 8.1 A simple correlation between property, cost, and performance for the success of a fuel cell device

proton conductivity, it often leads to increased water uptake and gas permeability. These changes can compromise the mechanical integrity of the membrane, thereby accelerating degradation and reducing overall system lifespan.

The operating conditions of a PEMFC device play a crucial role in determining its overall efficiency and durability. Several parameters—including the presence of impurities, cell temperature, operating voltage, current density, and fuel pressure—must be carefully controlled and optimized. Figure 8.1 illustrates a simplified correlation between material properties, cost, and performance for achieving an efficient fuel cell system. Furthermore, Figure 8.2 outlines the key technical parameters of PEMFC materials along with the industrial performance targets necessary for large-scale deployment. Impurities, in particular, have a profound impact on the long-term durability and functionality of PEMFCs. These contaminants can originate either from the fuel or the oxidant supply. For instance, hydrogen produced from natural gas or hydrocarbon sources may contain trace amounts of ammonia. The presence of ammonia can significantly impair the proton conductivity of the membrane by increasing ohmic resistance. As ammonia diffuses through the anode catalyst layer into the membrane, it can form ammonium ions (NH_4^+), which compete with protons (H^+) for transport sites, thus reducing the overall ionic conductivity and

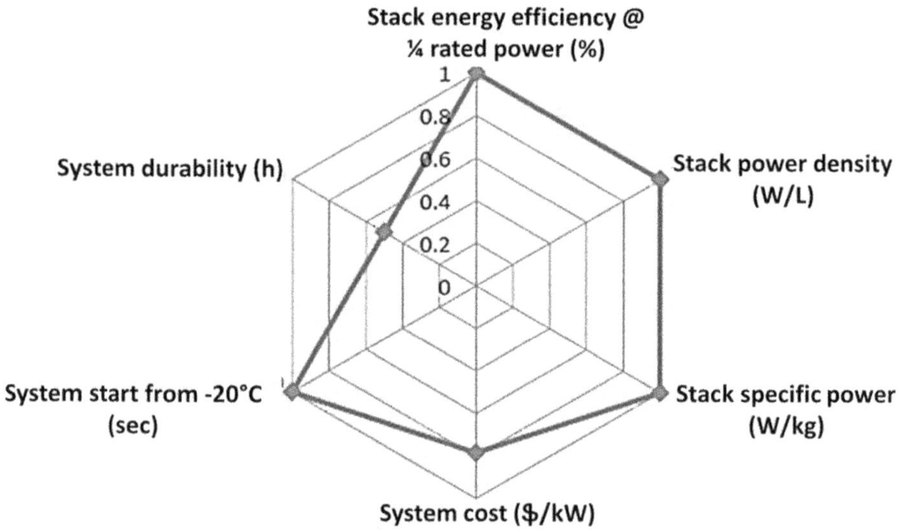

FIGURE 8.2 Technical parameters for PEMFC properties and industrialization targets: Blue line stands for current status as a fraction of targeted value [2] (Reprinted with permission)

performance of the PEM. Other common fuel-based contaminants include hydrogen sulfide (H_2S), hydrocarbons, formic acid, hydrogen cyanide, formaldehyde, and carbon monoxide. On the oxidant side, air can introduce sulfur oxides (SO_x), nitrogen oxides (NO_x), and particulate matter, all of which can adversely affect cell performance and catalyst activity. Furthermore, the method of hydrogen production from hydrocarbon fuels significantly influences the type and concentration of impurities present. The reforming and subsequent gas-cleaning steps determine the impurity profile. Notably, carbon dioxide (CO_2) can also lead to carbon monoxide (CO) generation via the reverse shift reaction, further complicating the purity of the fuel and exacerbating catalyst poisoning [1]. CO is the deadly impurity commonly known as anode catalyst poison forms strong bonds with metal particles, Pt by chemisorption at the metal surface. Chemisorption of CO affects hydrogen oxidation reactions by blocking the adsorption of hydrogen on to the active catalyst sites. This process adversely affects the electro-oxidation rates and increases the electrode overpotential resulting in undesirable performance loss. Moreover, SO_2 is a common contaminant obtained during the combustion of fossil fuels. The adverse effect of SO_2 is similar to H_2S and CO, where strong chemisorption on the Pt surface is responsible for decrease in the active sites for hydrogen adsorption resulting in the performance degradation with rise in SO_2 concentration.

8.3 MODES OF DEGRADATION OF PEM

8.3.1 Chemical Degradation

Membranes face rough environments both at the cathode and anode side of the MEA. The chemical degradation of PEM is mainly due to the peroxy and hydroperoxy attack of the radicals. The radical formation has been proposed by General Electric as mentioned in Eqns. 8.1–8.5. The peroxide radicals drastically attack the membrane. Based on the main chain of the polymer, the attack of radicals may proceed at the α-carbon side of the benzene ring or ether links or the branching sides. The hydrogen bearing side chains are also subjected to be attacked by the radicals more specifically at a higher temperature and low relative humidity.

$$H_2 \rightarrow 2H^{\bullet} \tag{8.1}$$

$$H^{\bullet} + O_2 \rightarrow HO_2^{\bullet} \tag{8.2}$$

$$HO_2^{\bullet} + H^{\bullet} \rightarrow H_2O_2 \tag{8.3}$$

$$H_2O_2 + M^{2+} \rightarrow M^{3+} + OH^{\bullet} + OH^{-} \tag{8.4}$$

$$OH^{\bullet} + H_2O_2 \rightarrow H_2O + HO_2^{\bullet} \tag{8.5}$$

The chemical degradation of the PSSA solution is studied by Laconti and coworkers [3]. Rapid degradation of this polymer has been evidenced in the presence of catalysts under the hydrogen atmosphere with a small amount of oxygen leading to the generation of chemically reactive H_2O_2. Linden et al. have proposed some equations that show the structural changes a polymer may undergo when it gets attacked by peroxides [4]. The degradation commences by the attack of the peroxide and abstraction of hydrogen from the polymer which proceeds further by progressive propagation steps. The proposed mechanism of degradation has been summarized in Eqn. 8.6–8.10 as follows.

$$PH + OH^{\bullet} \rightarrow P^{\bullet} + H_2O \tag{8.6}$$

$$P^{\bullet} + O_2 \rightarrow PO_2^{\bullet} \tag{8.7}$$

$$PO_2^{\bullet} + PH \rightarrow POOH + P^{\bullet} \tag{8.8}$$

$$PH + HO_2^{\bullet} \rightarrow P^{\bullet} + H_2O_2 \tag{8.9}$$

$$PH + HO_2^{\bullet} \rightarrow H^{\bullet} + HO_2^{\bullet} + P^{\bullet} \tag{8.10}$$

8.3.1.1 Chemical *Degradation* of Perfluorosulfonic Acid Membrane, Nafion

The chemical degradation of PFSA membranes involves radical-induced decomposition of the main chain of the polymer. The process of chemical degradation is yet to be clearly understood, however, the most accepted one is by the radical attack on the main and side chain of the Nafion membrane [5–8]. The electrochemical reaction of crossover gases on the catalyst surface may possibly leads to the formation of reactive oxygen species, however, they are of a short lifetime and almost incapable to diffuse significantly [9, 10]. Another possibility of radical formation is through the homolysis of H_2O_2 [11]. Additionally, the presence of iron and other metal ions generated from corrosion may also catalyze peroxide decomposition as depicted in Eqn. 8.11 [12]. The hydroxyl radical formed reacts with peroxide to generate hydroperoxyl as shown by Eqns. 8.12 and 8.13. A high concentration of hydroxyl radicals attacks the weak sites of Nafion, such as carboxylic acid ends, ether groups, tertiary carbons, and C-S bonds. Attack on carboxylic acid ends is also termed as unzipping of Nafion structure. Eqns. 8.14–8.16 explain the steps of the unzipping mechanism which leads to the degradation of the main chain. The second possible chemical idegradation can be by the attack of free radicals on C–S bonds of Nafion (Eqns. 8.17–8.19). At this step, the radical attack starts at the end of side-chain on C–S bond with progressive shortening of chain length and liberation of HF. At this stage, there is a possibility of generation of carboxyl groups which may further accelerate the process of unzipping on both sides of the main chain. The detection of respective radicals by UV-induced Fenton test supports the above-mentioned degradation mechanism [13]. This degradation leads to progressive losses of membrane weight as well as proton conductivity of the membranes. The third mechanism of degradation is based on the attack of the radicals on the ether groups of Nafion, i.e., the radical attack at O–C bonds. Studies propose that ether groups next to the C–S bond represent a suitable site for the radical attack in branched structure [14]. In this line, Danilczuk et al. have also demonstrated the fact that Nafion shows 20 times lower stability as compared to the short side chain-based ionomers [15]. Radicals may also attack the Nafion backbone at its tertiary carbon sides adjacent to the ether bonds. There exist two tertiary carbons in the Nafion structure; the attack on the tertiary carbon of the side-chain leads to side-chain degradation whereas the attack on the tertiary carbon at the main chain initiates chain scission. Due to the least thermodynamic stability, tertiary carbons are considered to be most prone to the radical attack [16, 17]. The unzipping mechanism is the most frequently discussed mode of degradation proposed by various research groups [18, 19]. A post-treated membrane having lower carboxylic ends observed to have better stability [15, 20]. Hence, it can be said that the membrane degradation is much dependent on the presence and nature of side chains. This is also expected that a membrane already suffering from degradation is more sensitive to the second attack. Figure 8.3 represents a radical attack on different positions of the chemical structure of Nafion.

$$H_2O_2 + M^{Z+} \rightarrow M^{(Z+1)+} + OH^{\bullet} + OH^- \tag{8.11}$$

$$OH^{\bullet} + H_2O_2 \rightarrow H_2O + HOO^{\bullet} \tag{8.12}$$

$$M^{(Z+1)+} + H_2O_2 \rightarrow M^{Z+} + HOO^{\bullet} + H^+ \tag{8.13}$$

$$R_f - CF_2COOH + OH^{\bullet} \rightarrow R_f - CF_2^{\bullet} + CO_2 + H_2O \tag{8.14}$$

$$R_f - CF_2^{\bullet} + OH^{\bullet} \rightarrow R_f - CF_2OH \rightarrow R_f - COF + HF \tag{8.15}$$

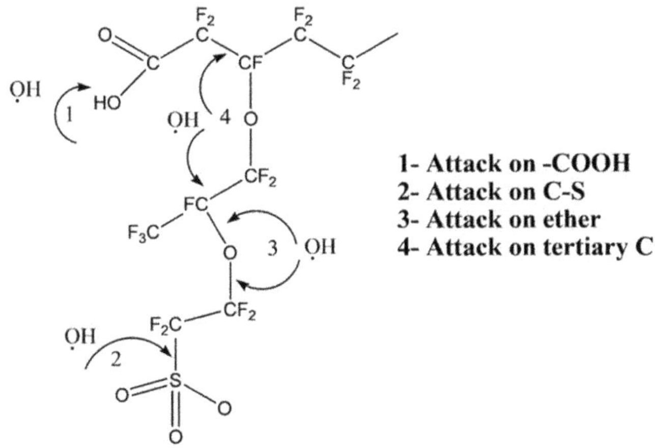

FIGURE 8.3 Radical attack on different positions of the chemical structure of Nafion (Adapted from [21])

$$R_f - COF + H_2O \rightarrow R_f - COOH + HF \tag{8.16}$$

$$R_f - CF_2SO_3H + OH^{\bullet} \rightarrow R_f - CF_2^{\bullet} + SO_3 + H_2O \tag{8.17}$$

$$R_f - CF_2 + OH^{\bullet} \rightarrow R_f - CF_2OH \rightarrow R_f - COF + HF \tag{8.18}$$

$$R_f - COF + HF \rightarrow R_f - COOH + HF \tag{8.19}$$

8.3.2 Thermal degradation

The major effect of temperature on Nafion is that with a rise in temperature the water content inside the membrane decreases which leads to dis-satisfactory cell performance. A noticeable change in the chemical structure of Nafion has been evidenced once the temperature reaches above 150°C. The high thermal stability of Nafion is attributed to the C-F bond strength and shielding effect induced by fluorines [22]. Above 200°C, the sulfonic acid groups start getting lost [23]. In another study, Samms et al. have proposed the thermal degradation behavior of Nafion [24]. When heating Nafion from 35 to 280°C under an inert atmosphere, it generates water and traces of oxides of sulfur and carbon. As the temperature is increased further, the evolution of other products has been evidenced with a progressive decrease in the water content in the degradation products. At higher temperature, the main degradation products are HF, SiF, and carbonyl fluorides [22]. Wilkie et al. have proposed degradation of Nafion by breakage of C-S bond with the generation of SO_2, OH, and carbon-based radicals which keep on degrading [22]. The literature results for the thermal degradation of Nafion and its nanocomposites reveal that Nafion degrades at a comparatively lower temperature than hydrides however the nanocomposites degrade faster at a higher temperature (450°C) [25].

The degradation around 300°C and 400°C is most critical since at this stage Nafion loses its ion exchange capability. They also noted that up to 400°C, Nafion releases a slight amount of SO_2, and the content of SO_2 emission increases rapidly when the temperature reaches about 480°C and attains a maximum emission at around 615°C.

8.3.3 Physical Degradation

Mechanical degradation of membranes can cause early life failure of thin polymeric membranes. For the PFSA membrane, the loss of fluoride is considered as a direct measure of membrane degradation initiated by mechanical failure [11]. Physical degradation in a membrane may create many defects such as pinholes, blisters, cracks, and punctures. Considerable changes have been observed in MEA after ten cycles of dynamic load [26]. The process of fabrication of membrane is a critical factor since the presence of undesirable foreign particles can be harmful and generate perforations. Careful attention is needed to be taken for the reactants inlets, seals, and edges, or any other potential points which may act as a weak region to initiate cracks or failure in the membrane. High temperature and pressure and low-hydration level can be a deadly combination accelerating the process of degradation by several orders of magnitude. Moreover, a non-uniform distribution of pressure between the MEA and bipolar plate induced during the process of fabrication or manufacturing may speed up membrane failure. The insufficient water content of the membrane also plays a detrimental effect on membrane life since a dry membrane becomes more brittle and fragile as expected. Another factor affecting the membrane life is the penetration of catalyst inside the membrane which may impart high local stress generation [23, 27]. Perforation can be deadly since it affects the cell voltage as well. Through the pinhole inside the membrane, the reactant gases pass through and react with a catalyst, which, in turn, counter balances electrochemical fuel cell reaction and a drop in cell voltage is observed. The heat of the reaction leads to membrane softening which allows more and more gas crossover and a drastic drop in cell performance is evidenced [23, 28]. The reinforcement of membrane is a common choice to improve mechanical strength. E-PTFE reinforced membranes are developed by Gore fuel cells which exhibit longer lifetime as compared to the non-reinforced one and more importantly the observation made there is for Gore membrane whose thickness is one-third as that of commercial Nafion membranes [29]. Figure 8.4 represents the physical degradation in the membrane and membrane electrode assembly.

Physical degradation within the PEM or MEA during fuel cell operation can lead to irreversible changes in membrane properties, directly impacting overall cell performance. Under operational conditions, the MEA is subjected to compressive forces between the bipolar plates, which can induce time-dependent deformation or creep in the membrane material. The progression of creep within the polymer matrix can be detrimental, resulting in membrane thinning and the formation of pinholes. Stucki et al. reported significant thinning of Nafion 117 membranes after approximately 15,000 hours of operation at 80°C, attributing the degradation to localized stress development on the membrane surface [32]. This suggests that mechanical stresses play a crucial role in accelerating membrane dissolution and physical failure.

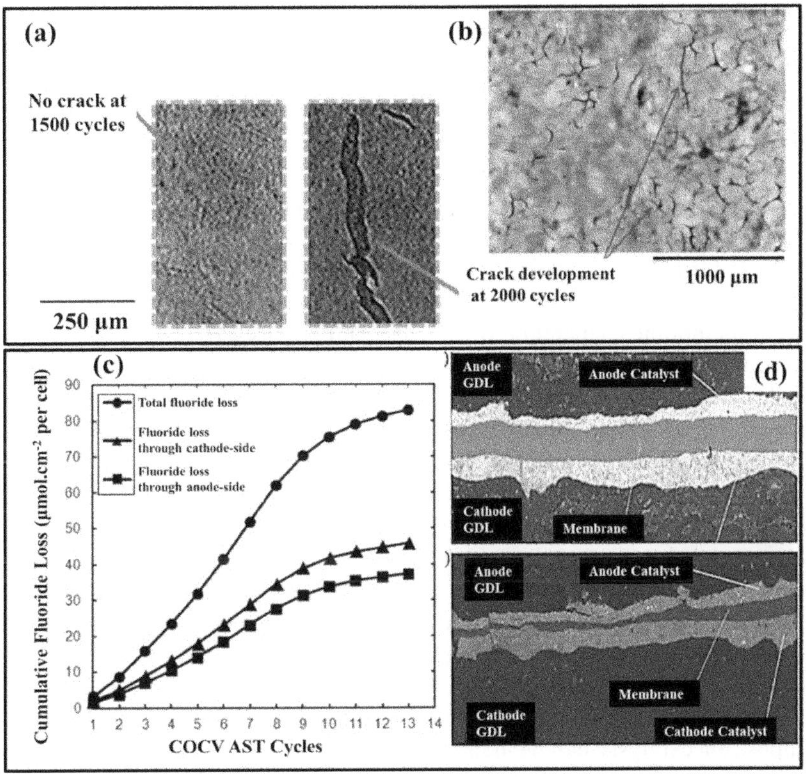

FIGURE 8.4 (a) Development of crack in PFSA membranes after 2000 cycles and (b) corresponding AFM topography [30], (c) cumulative fluoride loss with operating cycles and (d) degradation of membrane before and after operation showing membrane thinning when subjected to chemical and mechanical accelerated stress [31] (Reprinted with permission)

The extent of physical degradation and creep is also influenced by the membrane's structural characteristics. For example, sulfonated styrene–ethylene–butylene–styrene (SEBS) block copolymers with a low sulfonation degree (<6%) demonstrate enhanced creep resistance due to the intra-molecular hydrogen bonds between sulfonic acids, which reinforce the polymer structure [33]. Another commonly observed form of physical degradation is micro-crack formation, particularly at the edges of the flow channels where stress concentrations are highest [34]. These micro-cracks typically develop due to localized mechanical stretching, especially when pressure differentials exist across the membrane. Similarly, the interface between the reactive and non-reactive zones of the membrane is another vulnerable region. At this boundary, membrane failure can occur as a result of combined chemical and mechanical degradation, particularly through peroxide attacks, inadequate sealing of gaskets, and the formation of thermal hot spots [35].

The long-term stability of a PEMFC depends on both, temperature and level of hydration. The catastrophic failure may be delayed when fuel cells are operated under normal conditions, however, for Nafion, degradation initiates much early.

With an increase in temperature, the failure occurs much faster due to a decrease in mechanical properties under dry conditions. Under the wet conditions, the situation is more vulnerable since in the presence of water creep of the membrane increases. The structural or morphological changes are likely to have an impact on membrane durability although in an indirect way. Morphological changes in Nafion is reported by Yeo and Yeager on different forms on Nafion membranes [36]. They have classified the membranes in three forms, E-form or expanded form, N-form or normal form, and S-form or shrunken form. Membranes without thermal treatments are named as E-form, heat-treated membranes are named as N and S forms treated at 80 and 105°C, respectively. Zawodzinski et al. observe that the water uptake properties of Nafion membranes depend on the drying temperature [37, 38]. Membranes dried at higher temperatures undergo shrinkage in the ionic cluster as a result of which, uptake less water as compared to those dried at a relatively lower temperature.

Several strategies have been developed in recent years to soften degradation processes and extend the operational lifetime of PEMFCs. One key approach involves the reduction of metal-based impurities originating from various sources, including reactant gases, stack components, and bipolar/end plates. Minimizing these contaminants helps protect the membrane and catalyst layers from chemical degradation. Additionally, reducing gas permeability through the membrane, along with implementing efficient water and thermal management systems, plays a critical role in maintaining membrane integrity over prolonged operation. To counteract the harmful effects of radical-induced degradation—primarily from hydroxyl (•OH) and peroxide (H_2O_2) species, radical scavengers or inhibitors, as well as peroxide decomposition catalysts, can be incorporated directly into the membrane matrix. The inclusion of sacrificial materials that preferentially react with radicals before they reach the membrane backbone has also proven to be an effective protective strategy. Physical degradation mechanisms, such as pinhole formation, can be minimized through advanced diagnostic techniques, including in-situ temperature mapping, which enables the early detection of thermal hot spots within the cell. These localized regions are often precursors to mechanical failure.

8.4 EVALUATION OF MEMBRANE DEGRADATION

During the operation of a fuel cell device, chemical degradation within the membrane commences due to the reactive oxygen species commonly abbreviated as ROS. H_2O_2 is a common ROS generated by chemical and electrochemical pathways within the PEM which further produces reactive free radicals in the presence of metal ions via Fenton reaction [23, 39]. These reactive species are mainly composed of hydroxyl and hydroperoxyl radicals responsible for main and side chain degradations within PEM [18, 40]. As a consequence of this attack, the PEMFC suffers from severe stack failure. The suitable choice of radical scavengers will be worthy to minimize membrane failure. In the search for effective mitigation strategies, approaches are taken to induce radical scavengers. However, to effectively resolve the problem of degradation, it is essential to quantitatively analyze the rate of ROS generation during operation of the device. Detection of this reactive species is a difficult process since they have half-lives in the order of 10^{-9} s [41].

8.4.1 Steady state durability test

Steady state tests are conducted to find the durability or life span of PEMFCs under the operation condition of interest. The tests are run at a specific current density and the drop in the fuel cell performance is examined with operation time. Many research groups have tested to observe the voltage decay over time at a particular time of 4000–5000 h under current density of 0.4 A cm^{-2} [42–44]. The real-time degradation behavior of a PEM has been observed by inducing reactive oxygen [36]. The free radical scavengers, CeO_2 nanoparticles are introduced to catch the reactive oxygen species. The degradation pattern has been monitored by a fluorescent molecular probe in combination with in-situ fluorescence spectroscopy. CeO_2 acts as an effective radical scavenger that captures reactive oxygen for 7 h with an added advantage of eight-fold lowered reactive radical generation within the PEM.

As per the protocol of the steady state durability testing, the testing should be conducted without any break at a constant current density over a certain time period to observe the decay in output voltage, however it is very difficult from realistic point. In most of the studies the data acquisition takes place at regular intervals and the parameters are again carefully imposed to eliminate any reversible losses [45, 46]. The test parameters are also examined using various electrochemical characterization techniques such as impedance analysis, cyclic voltammetry, and polarization tests. The steady state test is a direct one however, it is not suitable for large-scale applications and also suffers from drawbacks of high cost and requirement of long time duration to complete the test.

8.4.2 In-situ accelerated stress test

In-situ accelerated test is done at an accelerated space to assess the degradation behavior of PEMFC where the device is subjected to simulated cycles. The assessment of degradation is done based on frequent start up and shut down, acceleration, sometimes idling, cold start, etc. Again the performance of device under acceleration and fuel starvation is also conducted to understand the response of the device under the real time scenario but at a faster space. The overall effect of these factors on the device performance is examined to understand and analyze the failure process. In situ PEM durability can be evaluated in both ways either by life-cycle test or by accelerated life-test. During the life cycle test, the PEMs are subjected to steady-state conditions since the conditions are easiest to apply. The general test conditions kept are at a temperature of 50°C–80°C and under the fully hydrated state. The OCV and reactant back pressure are variable as per targeted device applications. The failure of the membranes under the process is examined by AC impedance, microscopic studies; SEM, TEM, gas permeation, post-mortem investigation, etc. Stucki et al. have observed that the life span of a device with the Nafion 117 membrane reaches 15000 hours. when continuously operated at 80°C [32]. Ballard has established first, second and third-generation PEM for fuel cells named BAM1G, BAM2G, and BAM3G having chemical structures sulfonated polyphenylquinozaline, sulfonated polyphenylene oxide, and poly(trifluorostyrene sulfonic acid) [47]. During the in situ test, the durability of the membrane has been

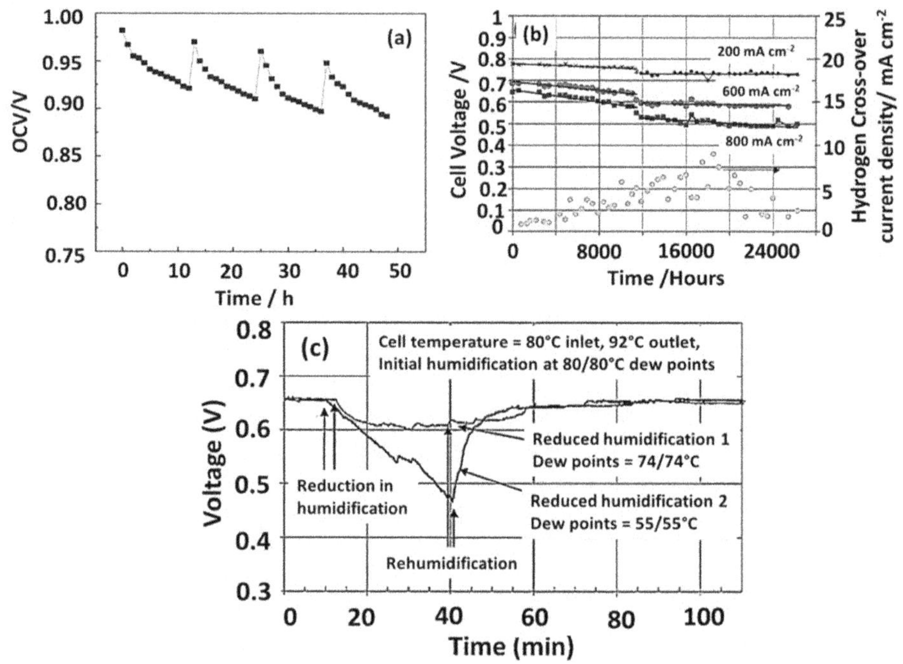

FIGURE 8.5 (a) OCV decay of Nafion single cell as a function of time [56], (b) Cell voltage and hydrogen cross-over current of a PEMFC operated at 70°C under 100% RH and ambient pressure with time during life test [45], and (c) Effect of humidity on PEMFC; current density: 1 A cm^{-2} [49] (Reprinted with permission)

evaluated either by fluoride-ion emission rate or gas cross-over rate. The chemical degradation of the PFSA membrane has been examined by Madden et al. by evaluating the release of fluoride ions during degradation of the membrane [48]. The study reveals that the release of fluoride ions lower as the operating temperature is kept low for adequate humidified membranes. Other interesting observations are the logarithmic decrement in the fluoride release rate with oxygen content and increment in fluoride emission rate with an increase in membrane thickness. This study claims that an accurate analysis of the rate of chemical degradation of the PFSA membrane can be obtained from fluoride emission rate if the testing has been performed in a controlled manner under uniform cell conditions. The main problem of the life-cycle test is that it's a time-taking process, may continue for as long as several thousands of hours. Accelerated test henceforth comes into the scenario to observe membrane durability faster [49]. Knights et al. have examined the aging and lifetime of PEMFC and DMFC at various conditions such as humidification level, temperature, and reactant flow rate (Figure 8.5).

8.4.3 Ex-situ accelerated stress test

During in situ test, degradation occurred at different components such as catalysts, membrane, bipolar plates, and gas diffusion layer, can easily be observed.

However, due to excessive time requirement and other constraints it is not very much fruitful to understand the major cause to device failure since various factors are operating simultaneously such as temperature, water content, and mechanical stress. Therefore, ex situ tests are performed where the test is conducted on a single component of the device at a time to understand the root cause of failure at every component. The ex situ test includes observations due to humidity, temperature, dry of wet condition of membranes, clamping stress effect, freeze-thaw cycling, etc.

An accelerated life test may be accompanied by a combination of four different parameters altogether or separately one by one. These are open-circuit voltage variation, reduced relative humidity, higher temperature, and different cycling studies such as relative humidity, voltage, temperature, and start-stop. Accelerated test speeds up the degradation of membrane and lifetime of the membranes get reduced to a few hundreds of hours [50]. The dryness of a membrane may lead to accelerated failure whereas the lifetime of the membranes can again be greatly affected by the humidification of reactant gases [45, 51, 52]. Figure 8.5 shows the variation of cell voltage and hydrogen cross-over current as a function of test time. OCV condition is an important factor controlling the peroxide generation. Generally, tests are performed at reduced humidity at OCV conditions to examine the effect of peroxide on the membrane lifetime [53]. Accelerated stress testing of a fuel cell has been studied by Tian et al. in recent work to observe the durability of fuel cell under actual operating conditions [54]. Cycling operation of fuel cell reveals that the thickness of the layer of the catalyst under study has been reduced progressively and the Pt particle size grows up compared to the fresh one. The electrochemical surface area degrades over time leading to performance decay of fuel cells. A combination of the chemical and mechanical accelerated stress tests has been conducted based on relative humidity cycling for PEMFC devices to observe the membrane durability in a recent study [55]. The study evidences the changes in membrane thickness and morphology as a function of field and drive cycle operation studies. The best benefit of accelerated testing is in its fast analysis of membrane failure with useful information on membrane durability.

8.5 TRANSPORT MANAGEMENT IN PEM

8.5.1 Electro osmotic effect and water management

The recipe for attaining efficient water management and improved fuel cell performance is better distribution of water within the PEMFC devices. The proton conductivity is greatly affected by the content of the water within the membrane. The water absorbed by sulfonated groups of Nafion enables free movement of protons by weakening the attraction force between the SO_3^- and H^+ groups. The more the water molecules are present, higher will be conductivity. Nonetheless, a huge amount of water will block the reactive sites of the adjacent electrodes and limit the access of reactant gases. Hence, an optimum balance between the two is essential to attain a satisfactory fuel cell performance. Water assists the formation of protons at the anode side, hydrates the PEM to facilitate the proton transport toward the cathode side and itself becomes a product due to the consumption of protons at the cathode side of the membrane electrode assembly. Water flux toward the anode under fuel cell operation may promote water flooding. An optimum balance of water content along the membrane electrode assembly is crucial to avoid flooding and maintain proper hydration levels. The water flux is denoted as electro-osmotic drag coefficient representing the number of water molecules transported per proton transport. For Nafion, an electro-osmotic drag coefficient of water is about 2.5. Now as 6 protons are generated by oxidation of one mole of methanol in the presence of water, hence 2.5 × 6 moles of water are to be dragged through the Nafion toward the cathode side. Again for the methanol-water mixed system, 16 moles of water molecules will be lost per mole of methanol [57]. Therefore, the content of water in the membrane affects the membrane-catalyst interface leading to the rupture of the membrane. Hence, proper humidification is to be maintained for the fuel and oxidant stream to avoid flooding along the membrane side and compensate the electro-osmotic drag coefficient at the anode side.

8.5.1.1 Water production and its effect on device

The effect of water production on the PEMFC performance is determined by the underlined electrochemical reaction that takes place therein. The byproduct water is formed on the surface of the catalyst particles by two gas-phase reactions namely oxygen dissociation and hydroperoxyl (-OOH) formation. The oxygen molecules get adsorbed on the platinum surface and undergo dissociation. The oxygen atoms then undergo a chemical reaction with the chemisorbed hydrogens to form water. Similarly, in the second reaction pathways, hydrogen atoms react with adsorbed oxygen at the surface of the catalyst and form OOH which in the subsequent step dissociates to produce chemisorbed O and hydroxyl (OH). This O and OH in the next step react with the chemisorbed hydrogen to generate water. Water generated at the catalyst surface makes it more exposed to deterioration as compared to the other side of the fuel cell. The PEMFC catalyst structure is mainly made of three major components namely, the electronically conducting component, i.e., Pt/C, an ionically conductive component, i.e., PFSA, and an insulating material, i.e., PTFE. Figure 8.6 represents the cross-sectional view of the membrane electrode assembly. These three components are distributed randomly as clusters bridging the membrane and current collector side. The gaseous reactants penetrate through the void present in those clusters. During the electrochemical process, these nanocomposites experience structural changes in Pt concentration and distribution. This structural change also reduces the active surface area of the catalyst governed by four main mechanisms say, Ostwald ripening, Pt crystal relocation and coalescence, detachment, and aggregation of Pt particles, dissolution and precipitation of Pt crystals in the membrane [44, 58]. As a consequence of catalyst deterioration, Pt particles also suffer from coarsening and aggregation. The potential cycling is the main reason for such deterioration, higher the number of cycles, more is the effect on catalyst particles. The stability of the Pt particle is also affected by corrosion of the carbon support during cell reactions [59]. To overcome the problem of catalyst deterioration, new alloys are developed to enhance the ORR activity, minimize the loss of catalyst particles from the reactive site, and reduce the corrosion of carbon support materials. Platinum-titanium

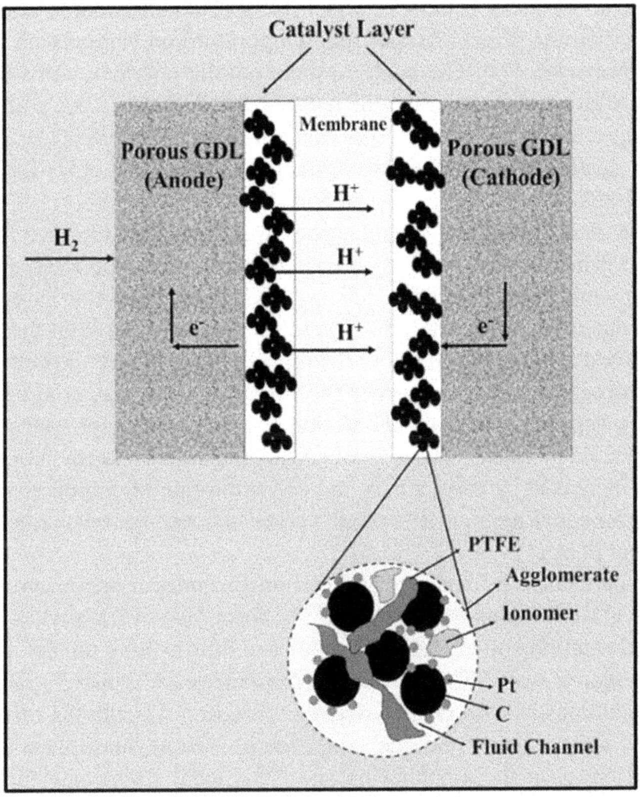

FIGURE 8.6 A cross-sectional view of the membrane electrode assembly (MEA)

and platinum-ruthenium/carbon are examples of suitable alloys with better catalyst stability under acidic environment. These alloys render the cells more CO tolerant with better ORR activity and current density. Again, the loss of Pt from the active catalyst site can be minimized by changing the catalyst design. Instead of bonding Pt with the carbon support, Pt can also be encapsulated inside fullerenes, called the metallofullerenes [60].

8.5.1.2 Water retention and accumulation

The excess of water in the PEMFC deteriorates the mechanical, chemical, and electrical properties of the overall device. Membrane swelling, corrosion, and formation of ice are the three major reasons to be considered in this respect. '*Membrane swelling*' occurs due to the absorption of water extending the volume of the polymer matrix. As the water molecules are absorbed in the membrane, ionic sites get surrounded by those water molecules while the polymer main chain resists further absorption of water molecules. The water inside the polymeric membrane enhances the internal pressure due to the elastic forces generated by the polymer matrix. The swelling of the polymeric membrane takes place in an anisotropic fashion as it creeps more toward one direction as compared to others. Again, the creep response of a membrane varies with the amount of water and surrounding temperature. Majsztrik et al. have studied the effect of water and temperature on Nafion with respect to its mechanical properties [61]. The study reveals that the interaction of water with sulfonic acid groups of Nafion makes the hydrophilic domains more stable. However, an increase in temperature breaks this domain apart. This complex effect plays a major role in deciding the number, strength, and flexibility of cross-links within the hydrophilic domains.

During the electrochemical cell reaction, hydrogen peroxides are formed. H_2O_2 facilitates the generation of hydroxyl (OH) and hydroperoxyl (OOH) species which degrade the membrane fast. These radicals lead to the reduction of active catalyst surface sites due to a significant reduction in the interface between the ionomer and catalyst particles. Relative humidity of the reactant gases also contributes to the generation of hydrogen peroxides. As per the research by Maass et al., the corrosion rate of carbon support is directly in relation to the molar water concentration [62]. For each molecule of water, one mole of CO_2 is generated. As the molar water concentration is increased by three times, the corrosion rate of carbon also found to be increased by an equal amount. Hence, the corrosion rate becomes maximum at the highest current density.

The damage caused by water accumulation becomes more severe as the water turns into ice at subfreezing temperatures. The formation of ice hinders the access of gases into the reaction site of the catalyst surface diminishing the cell performance. As per the research work [63], the cell performances get reduced gradually due to a fall in current density with progressive ice formation. During the formation of ice from water by freeze-thaw cycle, the expansion of volume takes place which further induces local stress in the neighboring site. This local stress may further degrade the membrane, catalyst, and GDL. In addition to the structural damage caused by the ice formation, the electrochemical surface area gets reduced further contributing to low cell performance. Ice formation may also promote the delamination of the catalyst layer. The water converted into ice accumulates either at the interface of the catalyst and the diffusion layer or catalyst and the membrane. Again, the formation of ice reduces the water content of the membrane hence, mobility of the protons gets reduced noticeably and a decrease in proton conductivity has been observed. Gallagher et al. have reported that the lowering of cell temperature to −25°C reduces the water content and hence, the proton transport coefficient of the membrane has been affected significantly [64]. Figure 8.7 represents the effect of water molar concentration, relative humidity, temperature and area resistance on the hydrogen crossover of Nafion membranes.

8.5.1.3 Water flooding

Flooding problem in PEMFC arises at high current densities when the rate of formation of water surpasses the rate of removal of the remaining water from the cell. Flooding is detected by monitoring the *in-situ* Ohmic resistance of the membrane. As reported, the difference in standard deviation on the electrochemical impedance measurement is an indication of water flooding [67]. Again as an effect of water flooding, a sudden decrement in voltage can also be observed. The constraints by residual water hinder the access of the reactant gases to the catalyst site.

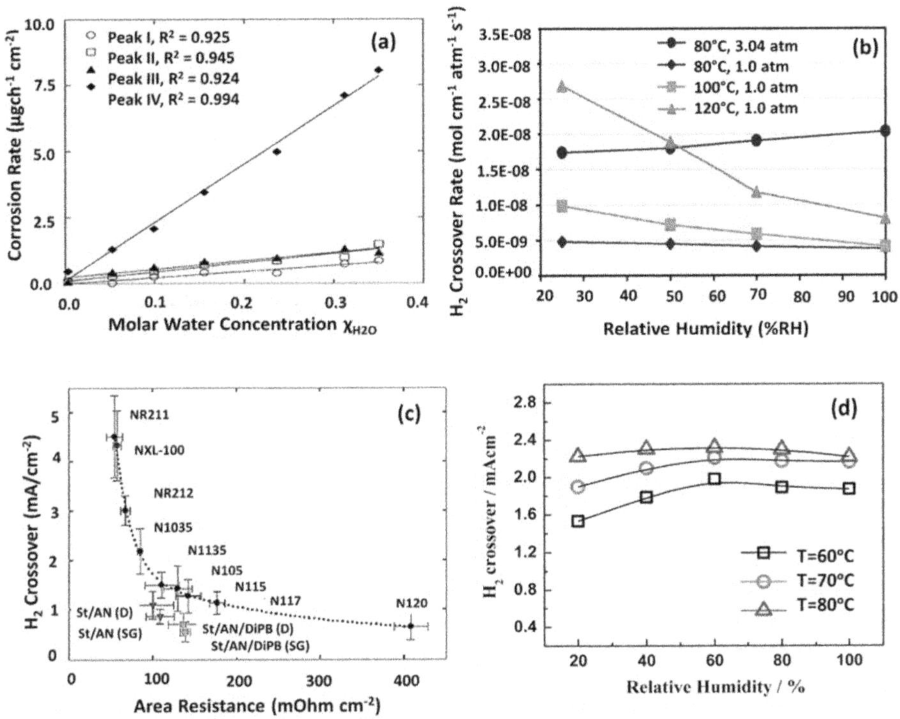

FIGURE 8.7 (a) Influence of molar water concentration on the rate of corrosion [62], (b) effect of relative humidity on the H_2 crossover rate [65], (c) H_2 crossover as a function of area resistance for different grades of Nafion [66], and (d) H_2 crossover with relative humidity and temperature [56] (Reprinted with permission)

A sharp drop in voltage occurs as soon as the limiting current density is reached also called cut-off density. Among all the detection techniques, optical visualization remains the most effective one to monitor the effect of water flooding in the cell. The main advantage of this technique is detecting the basic parameters that cause flooding and anticipating the action before the actual happening. Flooding not only effectively drops the device performance but also affects the durability of the cell. As a result of excessive water, the fuel cell is starved due to an increase in local electrode potential [68]. Reiser et al. have studied the effect of water flooding on the increment of cathode interfacial potential due to the cell starvation [69]. The high cathode potential induces carbon corrosion of the catalyst in no time as evidenced by a significant decrease in catalyst active reaction site and thinning of the cathode electrode with time. The flooding of water inside a MEA has been schematically represented in Figure 8.8.

8.5.2 Heat generation and management

In addition to proper water management, thermal management along the membrane electrode assembly side is also important for the proper functioning of the fuel cell

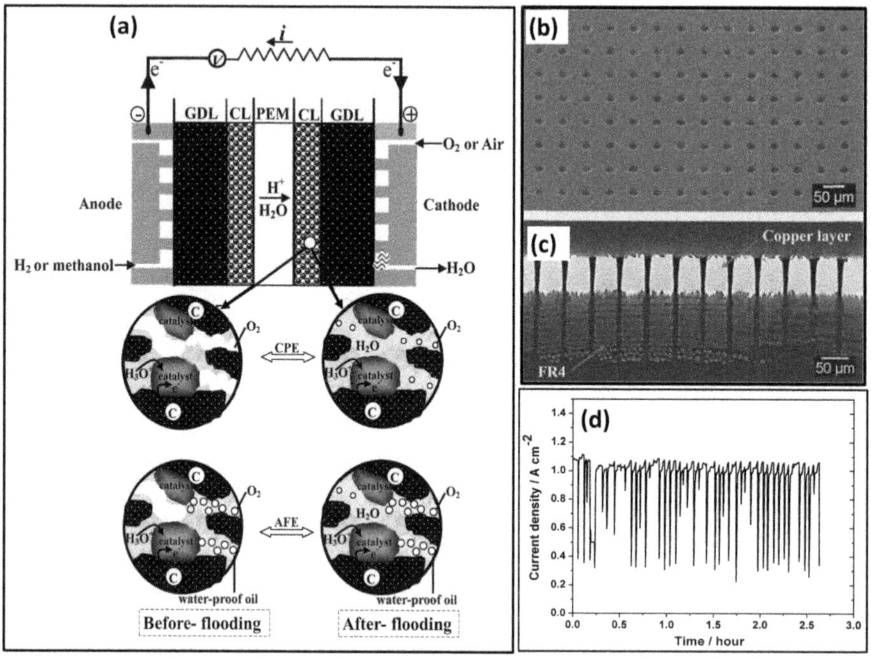

FIGURE 8.8 (a) Schematic representing condition of MEA flooding before and after flooding [70], (b) SEM of the capillaries on the copper layer, and (c) "xz" orthoslice prepared from tomogram made of a cross-section of a layer of PCB with capillary array [71], and (d) water flooding pattern in a device operating at a cell voltage of 0.6 V [72] (Reprinted with permission)

devices. The waste heat generation inside the PEMFC is dependent on its electric power output which may lead to a reduction of energy efficiency to about 50% [73]. To avoid the dehydration of the PEM, a low-temperature PEMFC is generally operated at the temperature range of 70°C–80°C which, however, gives the challenge to prevent flooding at the electrode side and hence voltage loss. Heat of reaction, Ohmic resistance, irreversible electrochemical reactions, water condensation, are the responsible factors that promote heat generation in the fuel cell devices [73, 74].

The huge amount of waste heat influences the overall cell performance that is majorly dependent on the operating temperature and temperature distribution inside the cell. The temperature management is again closely associated with the water management affecting device efficiency. The performance of the PEMFC is governed by two scenario, first the maintenance of a constant operating temperature, i.e., 80 °C and maintenance of isothermal condition over the complete cell area. The membrane hydration and proton conduction are again affected by the temperature. An excessive temperature would lead to the dehydration of the membrane leading to increase in resistivity and in other words reduction in proton conductivity. High temperature would result in fast membrane degradation reducing the durability of PEM and requirement for frequent replacement. However, low temperature leads to electrode flooding due to the formation of excess water and eventually the device would suffer oxygen starvation

and the mass concentration polarization which in turn will affect the reaction kinetics and reduction in proton conductivity. The uniform distribution of temperature over the cell is another important consideration. An uneven distribution would result in variation in relative humidity at places due to existence of different phases of water at places. The electrochemical reaction kinetics will also vary. The variation in reaction kinetic may create localized hot spots with improper utilization of reactant and catalyst affecting overall cell performance and harshly lowering the cell durability.

Various approaches have been developed for effective heat management to ensure optimal performance of PEMFCs. Air cooling is one of the most commonly employed techniques, particularly for low-power systems (<2 kW). In this method, heat is removed via airflow through the cooling channels embedded in the bipolar plates of the PEMFC stack [75]. A fan is typically used to inject pressurized air, aiding in the dissipation of excess heat and maintaining thermal stability. For higher power applications, liquid cooling systems are preferred due to their superior heat removal capacity. In such systems, a coolant, usually deionized water or a water–ethylene glycol mixture is circulated through dedicated channels in the bipolar plates using a pump. Liquid cooling is effective in maintaining uniform temperature distribution across the fuel cell stack, which is essential for consistent electrochemical performance and durability. However, despite its effectiveness, liquid cooling systems have certain limitations, including the risk of coolant leakage, potential corrosion of internal components, and the added operational and maintenance costs associated with the coolant circulation pump.

In recent studies, the heat management is done using phase change materials and metal hydrides [76]. Guo et al. developed a heat peak regulation system made of phase change material [77]. The phase change material is capable of exchanging heat with the liquid coolant and air conditioner refrigerant thereby successfully manages the heat dissipation. Researchers are trying to utilize the advantages of phase change material in combination with conventional cooling systems. With recent developments, the heat management strategies will not only manage the heat dissipation but also supply the heat requirement for startup in cold conditions. Chen et al. worked on the utilization of waste heat for cold start in fuel cell-operated electric vehicles [76].

8.5.3 Fuel Crossover and Oxidation

Fuel crossover or hydrogen crossover in the case of PEMFC is an unwanted diffusion of hydrogen from anode to the cathode side of the membrane. Hydrogen crossover adversely affects the fuel cell efficiency, depresses the cathode potential, and guides fast peroxide formation. The crossed over hydrogen reacts directly with the oxygen at the cathode side resulting in a reduction in cathode potential. Most importantly this direct reaction of oxygen with the hydrogen molecules forms peroxide radicals. This peroxide radicals attack the catalyst layer and membrane adversely, leading to severe membrane and catalyst layer degradation [78]. Moreover, this highly exothermic direct reaction between the hydrogen and oxygen creates hot points or pin-holes in the membranes abolishing MEA and raises a serious safety concern. Hydrogen crossover is a diffusion controlled process and is greatly affected by the membrane structure and operating condition of the fuel cell devices [79, 80]. Thin membrane operated at high temperature and pressure facilitates high hydrogen crossover. Using linear sweep voltammetry, the hydrogen crossover rates of various membranes such as Nafion, SPEEK, and Flemion, have been investigated [81–83]. The water content of the membrane and gas pressure also affect the hydrogen crossover rate in PEMFC [82, 84]. As discussed, hydrogen crossover leads to severe membrane degradation and also potentially affect the catalyst layer. Combined crossover of hydrogen and oxygen leads to combustion and also accelerating thermal degradation at localized regions. The sintering of catalyst takes place due to reduction of active surface area at the catalytic region. The mechanical degradation of the membrane is also accelerated due to migration of Pt through pores into the membrane leading to overall reduction in proton conduction [85]. A reduction in open circuit voltage and voltage efficiency are the observations.

Cheng et al. investigated the effect of temperature (80°C–120°C), back pressure (1.0–3.0 atm), and relative humidity (25%–100%) on the hydrogen crossover of PEMFC [65]. The model study indicates that diffusion of hydrogen molecules through PEM is the major parameter affecting the hydrogen crossover process. In this study, the hydrogen crossover is represented as an H_2 permeability coefficient. As per the study, at a given RH the H_2 permeability, coefficient increases with an increase in temperature and back pressure. However, the effect of RH on the permeability coefficient is complicated. Again, the study also reveals that the effect of temperature on the hydrogen crossover is much higher than that of pressure and relative humidity.

To assess fuel crossover in PEMFCs, two primary types of detection methods are employed, in situ and ex situ. Both electrochemical and non-electrochemical techniques are used to identify and evaluate the underlying failure mechanisms. In-situ measurements, conducted under actual fuel cell operating conditions, provide realistic and practical insights into crossover behavior. Electrochemical techniques commonly employed for crossover assessment include linear sweep voltammetry, potential step methods, and cyclic voltammetry. These methods help in quantifying hydrogen or methanol crossover by analyzing the resulting current response associated with the oxidation of the crossover species. Non-electrochemical approaches, however, involve analytical characterization tools such as mass spectrometry and gas chromatography. In mass spectrometric analysis, the spectrometer is connected to the outlet of MEA, and the concentration of hydrogen (or other crossover species) is measured. This measured concentration is then converted into a crossover rate over a defined time interval.

In gas chromatography-based analysis, one electrode is supplied with hydrogen while the opposite side is exposed to an inert gas. The gas chromatograph is connected to the outlet of the inert gas stream, allowing the detection and quantification of crossover species. This provides a direct evaluation of the crossover rate by analyzing the composition of gases diffused across the membrane [86].

However, since PEM are porous in nature, crossover of the fuel is unavoidable even when there is no or minimal defect in the device fabricated. Minor defects are always obvious during fabrication, installation and operational procedures. However, the crossover needs to be controlled or minimized to suppress the device failure by proper assessment and remedial action such that the PEMFC can operate efficiently for long service period.

8.5.4 CO POISONING AND MANAGEMENT

The presence of CO in the fuel gas leads to a common problem of CO poisoning in PEMFC devices. The preferential adsorption of CO over the platinum surface blocks the active sites of the catalyst particles and degrades the PEMFC performance. The Co poisoning can be mitigated by in depth understanding of poisoning mechanism. CO competes for hydrogen at the Pt catalyst side and adsorbs on its surface easily as compared to hydrogen due to higher heat of adsorption. Now due to the adsorption process the active side for the favorable fuel cell reaction is reduced with progressive formation of CO bed over the platinum surface. Below the reactions involved during adsorption and desorption and re-adsorption are mentioned in equations 8.20–8.22. Hence, there is a competition between two processes simultaneously.

$$CO + nPt \leftrightarrow Pt_n - CO_{ads} \tag{8.20}$$

$$2CO + 2(Pt - H_{ads}) \leftrightarrow 2(Pt - CO_{ads}) + H_2 \tag{8.21}$$

$$Pt - CO_{ads} + Pt - H2O_{ads} \leftrightarrow 2Pt + CO_2 = 2H^+ + 2e^- \tag{8.22}$$

The coverage of CO over catalyst surface increases with CO poisoning and at the threshold concentration electro-oxidation of CO occurs. For pure platinum, the electro-oxidation occurs at 0.6 V. CO covers about 98% of active sites of platinum at 1% CO/H_2 mixture. For controlling the CO poisoning and maintaining PEMFC durability, several approaches are taken. Strategies are taken to pre-treat reformate, in-situ techniques, on board removal of CO to avoid poisoning effect. Attempts have been taken to overcome the adverse effects of CO poisoning by (i) using platinum alloy as the catalyst material, (ii) elevating the fuel cell operating temperature, and (iii) introducing oxygen in the fuel gas stream [87, 88]. To eliminate the CO poisoning from the PEMFC devices, pure hydrogen is required to be used whereas in the case of DMFC devices the operating temperature should be elevated with the use of an effective electrocatalyst. The other factors affecting the fuel cell performance are membrane dehydration, mechanical stability, membrane conduction property, and stability in the oxidative environment. Under oxidizing conditions various free radicals, such as oxygen, hydroxide, and peroxides, are generated which attack the host polymer matrix leading to lowering of membrane performance due to the loss of side chain functionality. Hence, for the successful operation of fuel cell devices, thermal, mechanical, and oxidative stability of the membrane are of foremost consideration.

Current research trends focus extensively on understanding the detailed mechanisms of CO PEMFCs [89, 90]. One promising strategy involves the integration of CO sensors for real-time detection and control of poisoning effects, thereby enhancing system responsiveness and stability. Improving the CO tolerance of fuel cell systems is particularly advantageous, as it can significantly reduce the cost and complexity of hydrogen purification.

The development of CO-tolerant electrocatalysts has emerged as a key area of interest in recent studies. These catalysts aim to maintain high catalytic activity even in the presence of trace amounts of CO. In a novel approach, the use of hexavalent chromium

TABLE 8.1
Technical parameters for PEMFC devices

Parameters	Unit	Value	Reference
Specific power	$kW\ kg^{-1}$	400–1000	[92]
Electrical capacity	kW	0.75–2	[93–95]
Thermal capacity	kW	0.75–2	[93–95]
Electrical efficiency	%	35–39	[93–95]
Thermal efficiency	%	55	[93–95]
Degradation rate	Per year (%)	1	[93–95]
Pt loading	$Mg\ cm^{-2}$	0.2–0.8	[96]
CO tolerance	ppm	<50	[96]
Power density	$kW\ L^{-1}$	600–2000	[92]
Cold start?	-	Yes	[96]
Water management	-	Complex	[96]
Applications		Residential	[93–95]

(Cr(VI)) for the selective removal of CO has been explored, wherein Cr(VI) is reduced to Cr(III) via a redox reaction with CO. This process not only mitigates CO poisoning but also simultaneously eliminates the toxic Cr(VI) species, offering a dual environmental and operational benefit [91]. CO poisoning remains a critical challenge, as it significantly impacts both the durability and overall performance of PEMFCs by deactivating active sites on the anode catalyst. To address this, researchers are increasingly employing computational modeling and simulations to gain deeper insight into the adsorption and reaction dynamics of CO on catalyst surfaces.

Advancements in CO mitigation strategies and improved CO-tolerant materials are expected to substantially lower the burden and cost of hydrogen purification. Table 8.1 summarizes the key technical parameters necessary for the successful operation of PEMFC devices.

8.6 CHAPTER SUMMARY

The durability of the low temperature fuel cell is of major concern to achieve long term performance of the PEM without need of frequent change. Various factors contribute to the degradation of PEM such as membrane thinning, fuel crossover, excessive water flooding, and CO poisoning. The degradation of PEM again takes place by different routes such as chemical degradation, thermal degradation and physical degradation. To evaluate the performance and durability of PEM during prolonged operation, various tests are performed. The tests are performed either under steady state condition or under accelerated condition. The accelerated tests can again be conducted in-site or ex-situ mode. The durability of PEM is again regulated by optimized water and thermal management inside the device. CO poisoning, water flooding, fuel crossover, etc., promote serious problems in the device leading to the ultimate failure in the device performance. In this chapter, we have discussed the major factors affecting the degradation of PEM, the associated challenges to overcome the thermal and water management in PEM and the possible remedies to resolve these issues have been included.

REFERENCES

[1] Gu T, Lee W-K, Zee JW Van. Quantifying the 'reverse water gas shift' reaction inside a PEM fuel cell. *Appl Catal B* 2005;56:43–50. https://doi.org/10.1016/j.apcatb.2004.08.016.
[2] Wang C, Wang S, Peng L, Zhang J, Shao Z, Huang J, et al. Recent progress on the key materials and components for proton exchange membrane fuel cells in vehicle applications. *Energies (Basel)* 2016;9:603–41. https://doi.org/10.3390/en9080603.
[3] LaConti AB, Hamdan M, McDonald RC. *Handbook of fuel cells–fundamentals, technology and applications.* vol. 3. Wiley; 2003.
[4] Lindén LÅ, Rabek JF, Kaczmarek H, Kaminska A, Scoponi M. Photooxidative degradation of polymers by HOxxx and HO2xxx radicals generated during the photolysis of H2O2, FeCl3, and Fenton reagents. *Coord Chem Rev* 1993;125:195–217. https://doi.org/10.1016/0010-8545(93)85018-Y.
[5] Gubler L, Koppenol WH. Kinetic simulation of the chemical stabilization mechanism in fuel cell membranes using cerium and manganese redox couples. *J Electrochem Soc* 2011;159:B211–8. https://doi.org/10.1149/2.075202jes.
[6] Nosaka Y, Ohtaka K, Ohguri N, Nosaka AY. Detection of OH radicals generated in polymer membranes of PEFC. *ECS Trans* 2010;33:899–905. https://doi.org/10.1149/1.3484583.
[7] Ohguri N, Nosaka AY, Nosaka Y. Detection of OH radicals formed at PEFC electrodes by means of a fluorescence probe. *Electrochem Solid-State Lett* 2009;12:B94–6. https://doi.org/10.1149/1.3106184.
[8] Ohguri N, Nosaka AY, Nosaka Y. Detection of OH radicals as the effect of Pt particles in the membrane of polymer electrolyte fuel cells. *J Power Sources* 2010;195:4647–52. https://doi.org/10.1016/j.jpowsour.2010.02.010.
[9] Endoh E, Terazono S, Widjaja H, Takimoto Y. Degradation study of MEA for PEMFCs under low humidity conditions. *Electrochem Solid-State Lett* 2004;7:A209–11. https://doi.org/10.1149/1.1739314.
[10] Gubler L, Dockheer SM, Koppenol WH. Radical (HO , H • and HOO •) Formation and ionomer degradation in polymer electrolyte fuel cells. *J Electrochem Soc* 2011;158: B755–69. https://doi.org/10.1149/1.3581040.
[11] Pozio A, Silva RF, De Francesco M, Giorgi L. Nafion degradation in PEFCs from end plate iron contamination. *Electrochim Acta* 2003;48:1543–9. https://doi.org/10.1016/S0013-4686(03)00026-4.
[12] Danilczuk M, Coms FD, Schlick S. Visualizing chemical reactions and crossover processes in a fuel cell inserted in the ESR resonator: Detection by spin trapping of oxygen radicals, nafion-derived fragments, and hydrogen and deuterium atoms. *J Phys Chem B* 2009;113:8031–42. https://doi.org/10.1021/jp901597f.
[13] Bosnjakovic A, Kadirov MK, Schlick S. Using ESR spectroscopy to study radical intermediates in proton-exchange membranes exposed to oxygen radicals. *Res Chem Intermed* 2007;33:677–87. https://doi.org/10.1163/156856707782169372.
[14] Dreizler AM, Roduner E. Reaction kinetics of hydroxyl radicals with model compounds of fuel cell polymer membranes. *Fuel Cells* 2012;12:132–40. https://doi.org/10.1002/fuce.201100157.
[15] Danilczuk M, Perkowski AJ, Schlick S. Ranking the stability of perfluorinated membranes used in fuel cells to attack by hydroxyl radicals and the effect of Ce(III): A competitive kinetics approach based on spin trapping ESR. *Macromolecules* 2010;43:3352–8. https://doi.org/10.1021/ma1001386.
[16] Coms FD. The chemistry of fuel cell membrane chemical degradation. *ECS Trans* 2008;16:235–55. https://doi.org/10.1149/1.2981859.
[17] Tokumasu T, Ogawa I, Koyama M, Ishimoto T, Miyamoto A. A DFT study of bond dissociation trends of perfluorosulfonic acid membrane. *J Electrochem Soc* 2011;158:B175–9. https://doi.org/10.1149/1.3518421.

[18] Curtin DE, Lousenberg RD, Henry TJ, Tangeman PC, Tisack ME. Advanced materials for improved PEMFC performance and life. *J Power Sources* 2004;131:41–8. https://doi.org/10.1016/j.jpowsour.2004.01.023.
[19] Healy J, Hayden C, Xie T, Olson K, Waldo R, Brundage M, et al. Aspects of the chemical degradation of PFSA ionomers used in PEM fuel cells. *Fuel Cells* 2005;5:302–8. https://doi.org/10.1002/fuce.200400050.
[20] Zhou C, Guerra MA, Qiu Z-M, Zawodzinski, Thomas A., Schiraldi DA. Chemical durability studies of perfluorinated sulfonic acid polymers and model compounds under mimic fuel cell conditions. *Macromolecules* 2007;40:8695–707. https://doi.org/10.1021/ma071603z.
[21] Zatoń M, Rozière J, Jones DJ. Current understanding of chemical degradation mechanisms of perfluorosulfonic acid membranes and their mitigation strategies: A review. *Sustain Energy Fuels* 2017;1:409–38. https://doi.org/10.1039/C7SE00038C.
[22] Wilkie CA, Thomsen JR, Mittleman ML. Interaction of poly(methyl methacrylate) and nafions. *J Appl Polym Sci* 1991;42:901–9. https://doi.org/10.1002/app.1991.070420404.
[23] Reitz W. Handbook of fuel cells: Fundamentals, technology, and applications, (Volume 1) W. Vielstich, A. Lamm, and H. A. Gasteiger (editors). *Mater Manuf Processes* 2007; 22:788–788. https://doi.org/10.1080/10426910701416310.
[24] Samms SR, Wasmus S, Savinell RF. Thermal stability of nafion® in simulated fuel cell environments. *J Electrochem Soc* 1996;143:1498–504. https://doi.org/10.1149/1.1836669.
[25] Deng Q, Wilkie CA, Moore RB, Mauritz KA. TGA–FTi.r. investigation of the thermal degradation of Nafion® and Nafion®/[silicon oxide]-based nanocomposites. *Polymer (Guildf)* 1998;39:5961–72. https://doi.org/10.1016/S0032-3861(98)00055-X.
[26] Wang G, Huang F, Yu Y, Wen S, Tu Z. Degradation behavior of a proton exchange membrane fuel cell stack under dynamic cycles between idling and rated condition. *Int J Hydrogen Energy* 2018;43:4471–81. https://doi.org/10.1016/j.ijhydene.2018.01.020.
[27] Huang C, Seng Tan K, Lin J, Lee Tan K. XRD and XPS analysis of the degradation of the polymer electrolyte in H2–O2 fuel cell. *Chem Phys Lett* 2003;371:80–5. https://doi.org/10.1016/S0009-2614(03)00259-8.
[28] Yu J, Matsuura T, Yoshikawa Y, Islam MN, Hori M. In Situ Analysis of Performance Degradation of a PEMFC under Nonsaturated Humidification. *Electrochem Solid-State Lett* 2005;8:A156–8. https://doi.org/10.1149/1.1854781.
[29] Liu W, Ruth K, Rusch G. The membrane durability in PEM fuel cells. *J New Mater Electrochem Syst* 2001;4:227–32.
[30] Singh Y, White RT, Najm M, Haddow T, Pan V, Orfino FP, et al. Tracking the evolution of mechanical degradation in fuel cell membranes using 4D in situ visualization. *J Power Sources* 2019;412:224–37. https://doi.org/10.1016/j.jpowsour.2018.11.049.
[31] Lim C, Ghassemzadeh L, Van Hove F, Lauritzen M, Kolodziej J, Wang GG, et al. Membrane degradation during combined chemical and mechanical accelerated stress testing of polymer electrolyte fuel cells. *J Power Sources* 2014;257:102–10. https://doi.org/10.1016/j.jpowsour.2014.01.106.
[32] Stucki S, Scherer GG, Schlagowski S, Fischer E. PEM water electrolysers: Evidence for membrane failure in 100kW demonstration plants. *J Appl Electrochem* 1998;28:1041–9. https://doi.org/10.1023/A:1003477305336.
[33] Blackwell RI, Mauritz KA. Mechanical creep and recovery of poly(styrene-b-ethylene/butylene-b-styrene) (SEBS), sulfonated SEBS (sSEBS), and sSEBS/silicate nanostructured materials. *Polym Adv Technol* 2005;16:212–20. https://doi.org/10.1002/pat.574.
[34] Kundu S, Simon LC, Fowler M, Grot S. Mechanical properties of NafionTM electrolyte membranes under hydrated conditions. *Polymer (Guildf)* 2005;46:11707–15. https://doi.org/10.1016/j.polymer.2005.09.059.
[35] Wang H, Capuano GA. Behavior of raipore radiation-grafted polymer membranes in H 2/O 2 fuel cells. *J Electrochem Soc* 1998;145:780–4. https://doi.org/10.1149/1.1838345.

[36] Sone Y, Ekdunge P, Simonsson D. Proton conductivity of nafion 117 as measured by a four-electrode AC impedance method. *J Electrochem Soc* 1996;143:1254–9. https://doi.org/10.1149/1.1836625.
[37] Zawodzinski TA, Springer TE, Davey J, Jestel R, Lopez C, Valerio J, et al. A comparative study of water uptake by and transport through ionomeric fuel cell membranes. *J Electrochem Soc* 1993;140:1981–5. https://doi.org/10.1149/1.2220749.
[38] Zawodzinski TA, Derouin C, Radzinski S, Sherman RJ, Smith VT, Springer TE, et al. Water uptake by and transport through nafion® 117 membranes. *J Electrochem Soc* 1993;140:1041–7. https://doi.org/10.1149/1.2056194.
[39] Woodmansee AN, Imlay JA. Reduced flavins promote oxidative DNA damage in non-respiring escherichia coli by delivering electrons to intracellular free iron. *J Biol Chem* 2002;277:34055–66. https://doi.org/10.1074/jbc.M203977200.
[40] Schiraldi DA. Perfluorinated polymer electrolyte membrane durability. *J Macromol Sci, Part C: Polym Rev* 2006;46:315–27. https://doi.org/10.1080/15583720600796458.
[41] SIES H. Strategies of antioxidant defense. *Eur J Biochem* 1993;215:213–9. https://doi.org/10.1111/j.1432-1033.1993.tb18025.x.
[42] Pedersen LS, Nielsen HP, Kiil S, Hansen LA, Dam-Johansen K, Kildsig F, et al. Full-scale co-firing of straw and coal. *Fuel* 1996;75:1584–90. https://doi.org/10.1016/0016-2361(96)82642-1.
[43] Sishtla C, Koncar G, Platon R, Gamburzev S, Appleby AJ, Velev OA. Performance and endurance of a PEMFC operated with synthetic reformate fuel feed. *J Power Sources* 1998;71:249–55. https://doi.org/10.1016/S0378-7753(97)02719-5.
[44] Cheng X, Chen L, Peng C, Chen Z, Zhang Y, Fan Q. Catalyst microstructure examination of PEMFC membrane electrode assemblies vs. Time. *J Electrochem Soc* 2004;151:A48–52. https://doi.org/10.1149/1.1625944.
[45] Cleghorn SJC, Mayfield DK, Moore DA, Moore JC, Rusch G, Sherman TW, et al. A polymer electrolyte fuel cell life test: 3 years of continuous operation. *J Power Sources* 2006;158:446–54. https://doi.org/10.1016/j.jpowsour.2005.09.062.
[46] Verhage AJL, Coolegem JF, Mulder MJJ, Yildirim MH, de Bruijn FA. 30,000 h operation of a 70 kW stationary PEM fuel cell system using hydrogen from a chlorine factory. *Int J Hydrogen Energy* 2013;38:4714–24. https://doi.org/10.1016/j.ijhydene.2013.01.152.
[47] Steck A, Stone C. Development of the BAM membrane for fuel cell applications. *Proceedings of the 2nd international symposium on new material for fuel cell and modern battery systems*, Montreal, Canada; 1997, p. 792–807.
[48] Madden T, Weiss D, Cipollini N, Condit D, Gummalla M, Burlatsky S, et al. Degradation of polymer-electrolyte membranes in fuel cells. *J Electrochem Soc* 2009;156:B657–62. https://doi.org/10.1149/1.3095466.
[49] Knights SD, Colbow KM, St-Pierre J, Wilkinson DP. Aging mechanisms and lifetime of PEFC and DMFC. *J Power Sources* 2004;127:127–34. https://doi.org/10.1016/j.jpowsour.2003.09.033.
[50] Lakshmanan B, Huang W, Olmeijer D, Weidner JW. Polyetheretherketone Membranes for Elevated Temperature PEMFCs. *Electrochem Solid-State Lett* 2003;6:A282–5. https://doi.org/10.1149/1.1619647.
[51] Inaba M, Yamada H, Umebayashi R, Sugishita M, Tasaka A. Membrane degradation in polymer electrolyte fuel cells under low humidification conditions. *Electrochemistry* 2007;75:207–12. https://doi.org/10.5796/electrochemistry.75.207.
[52] Luo Z, Li D, Tang H, Pan M, Ruan R. Degradation behavior of membrane–electrode-assembly materials in 10-cell PEMFC stack. *Int J Hydrogen Energy* 2006;31:1831–7. https://doi.org/10.1016/j.ijhydene.2006.02.029.
[53] *2005 DOE hydrogen program review.* 2005:23–6. United States Department of Energy (DOE).

[54] Tian T, Tang J, Chen Y, Tan J, Li S, Pan M. Study on accelerated stress test for fuel cell lifetime. *Int J Electrochem Sci* 2018;13:2022–32. https://doi.org/10.20964/2018.02.69.
[55] Mukundan R, Baker AM, Kusoglu A, Beattie P, Knights S, Weber AZ, et al. Membrane accelerated stress test development for polymer electrolyte fuel cell durability validated using field and drive cycle testing. *J Electrochem Soc* 2018;165:F3085–93. https://doi.org/10.1149/2.0101806jes.
[56] Zhang H, Li J, Tang H, Lin Y, Pan M. Hydrogen crossover through perfluorosulfonic acid membranes with variable side chains and its influence in fuel cell lifetime. *Int J Hydrogen Energy* 2014;39:15989–95. https://doi.org/10.1016/j.ijhydene.2014.01.076.
[57] Ise M, Kreuer KD, Maier J. Electroosmotic drag in polymer electrolyte membranes: An electrophoretic NMR study. *Solid State Ion* 1999;125:213–23. https://doi.org/10.1016/S0167-2738(99)00178-2.
[58] Bi W, Fuller TF. Modeling of PEM fuel cell Pt/C catalyst degradation. *J Power Sources* 2008;178:188–96. https://doi.org/10.1016/j.jpowsour.2007.12.007.
[59] Ball SC, Hudson SL, Theobald B, Thompsett D. The effect of dynamic and steady state voltage excursions on the stability of carbon supported Pt and PtCo catalysts. *ECS Trans* 2006;3:595–605. https://doi.org/10.1149/1.2356180.
[60] Gabriel MA, Deutsch T, Franco AA. Fullerene-based materials as catalysts for fuel cells. *ECS Trans* 2010;25:1–6. https://doi.org/10.1149/1.3301100.
[61] Majsztrik PW, Bocarsly AB, Benziger JB. Viscoelastic response of nafion. Effects of temperature and hydration on tensile creep. *Macromolecules* 2008;41:9849–62. https://doi.org/10.1021/ma801811m.
[62] Maass S, Finsterwalder F, Frank G, Hartmann R, Merten C. Carbon support oxidation in PEM fuel cell cathodes. *J Power Sources* 2008;176:444–51. https://doi.org/10.1016/j.jpowsour.2007.08.053.
[63] Lee S, Roberts J, Li J, Ye S. Characterization of the catalyst layer in a PEMFC during subzero operation. *ECS Trans* 2008;12:13–9. https://doi.org/10.1149/1.2921528.
[64] Gallagher KG, Pivovar BS, Fuller TF. Electro-osmosis and water uptake in polymer electrolytes in equilibrium with water vapor at low temperatures. *J Electrochem Soc* 2009;156:B330–8. https://doi.org/10.1149/1.3046150.
[65] Cheng X, Zhang J, Tang Y, Song C, Shen J, Song D, et al. Hydrogen crossover in high-temperature PEM fuel cells. *J Power Sources* 2007;167:25–31. https://doi.org/10.1016/j.jpowsour.2007.02.027.
[66] Albert A, Barnett AO, Thomassen MS, Schmidt TJ, Gubler L. Radiation-grafted polymer electrolyte membranes for water electrolysis cells: Evaluation of key membrane properties. ACS *Appl Mater Interfaces* 2015;7:22203–12. https://doi.org/10.1021/acsami.5b04618.
[67] Roy S, Orazem ME. Stochastic analysis of flooding in PEM fuel cells by electrochemical impedance spectroscopy. *ECS Trans* 2007;11:485–95. https://doi.org/10.1149/1.2780962.
[68] Halalay IC, Swathirajan S, Merzougui B, Balogh MP, Garabedian GC, Carpenter MK. Anode materials for mitigating hydrogen starvation effects in PEM fuel cells. *J Electrochem Soc* 2011;158:B313–21. https://doi.org/10.1149/1.3530796.
[69] Reiser CA, Bregoli L, Patterson TW, Yi JS, Yang JD, Perry ML, et al. A reverse-current decay mechanism for fuel cells. *Electrochemical and Solid-State Letters* 2005;8:A273–6. https://doi.org/10.1149/1.1896466.
[70] Ji MB, Wei ZD, Chen SG, Li L. A Novel antiflooding electrode for proton exchange membrane fuel cells. *J Phys Chem C* 2009;113:765–71. https://doi.org/10.1021/jp807773m.
[71] Cho JIS, Neville TP, Trogadas P, Bailey J, Shearing P, Brett DJL, et al. Capillaries for water management in polymer electrolyte membrane fuel cells. *Int J Hydrogen Energy* 2018;43:21949–58. https://doi.org/10.1016/j.ijhydene.2018.10.030.

[72] Ji M, Wei Z. A review of water management in polymer electrolyte membrane fuel cells. *Energies (Basel)* 2009;2:1057–106. https://doi.org/10.3390/en20401057.
[73] Goebel SG Google Patents. Evaporative cooled fuel cell. US Patent 6,960,404, 2005.
[74] Kandlikar SG, Lu Z. Thermal management issues in a PEMFC stack – A brief review of current status. *Appl Therm Eng* 2009;29:1276–80. https://doi.org/10.1016/j.applthermaleng.2008.05.009.
[75] Amirfazli A, Asghari S, Sarraf M. An investigation into the effect of manifold geometry on uniformity of temperature distribution in a PEMFC stack. *Energy* 2018;145:141–51. https://doi.org/10.1016/j.energy.2017.12.124.
[76] Chen Q, Zhang G, Zhang X, Sun C, Jiao K, Wang Y. Thermal management of polymer electrolyte membrane fuel cells: A review of cooling methods, material properties, and durability. *Appl Energy* 2021;286:116496–514. https://doi.org/10.1016/j.apenergy.2021.116496.
[77] Guo J, Li Z, Wei L, Fangming J. A novel thermal management system with a heat-peak regulator for fuel cell vehicles. *J Clean Prod* 2023;414:137712. https://doi.org/10.1016/j.jclepro.2023.137712.
[78] Collier A, Wang H, Ziyuan X, Zhang J, Wilkinson D. Degradation of polymer electrolyte membranes. *Int J Hydrogen Energy* 2006;31:1838–54. https://doi.org/10.1016/j.ijhydene.2006.05.006.
[79] Song Y, Fenton JM, Kunz HR, Bonville LJ, Williams M V. High-performance PEMFCs at elevated temperatures using nafion 112 membranes. *J Electrochem Soc* 2005;152:A539–44. https://doi.org/10.1149/1.1855871.
[80] Yeager HL, Steck A. Cation and water diffusion in nafion ion exchange membranes: Influence of polymer structure. *J Electrochem Soc* 1981;128:1880–4. https://doi.org/10.1149/1.2127757.
[81] Jiang R, Kunz HR, Fenton JM. Investigation of membrane property and fuel cell behavior with sulfonated poly(ether ether ketone) electrolyte: Temperature and relative humidity effects. *J Power Sources* 2005;150:120–8. https://doi.org/10.1016/j.jpowsour.2005.03.180.
[82] Sakai T, Takenaka H, Torikai E. Gas diffusion in the dried and hydrated nafions. *J Electrochem Soc* 1986;133:88–92. https://doi.org/10.1149/1.2108551.
[83] Yoshida N, Ishisaki T, Watakabe A, Yoshitake M. Characterization of Flemion® membranes for PEFC. *Electrochim Acta* 1998;43:3749–54. https://doi.org/10.1016/S0013-4686(98)00133-9.
[84] Inaba M, Kinumoto T, Kiriake M, Umebayashi R, Tasaka A, Ogumi Z. Gas crossover and membrane degradation in polymer electrolyte fuel cells. *Electrochim Acta* 2006;51:5746–53. https://doi.org/10.1016/j.electacta.2006.03.008.
[85] Jung A, Oh J, Han K, Kim MS. An experimental study on the hydrogen crossover in polymer electrolyte membrane fuel cells for various current densities. *Appl Energy* 2016;175:212–7. https://doi.org/10.1016/j.apenergy.2016.05.016.
[86] Tang Q, Li B, Yang D, Ming P, Zhang C, Wang Y. Review of hydrogen crossover through the polymer electrolyte membrane. *Int J Hydrogen Energy* 2021;46:22040–61. https://doi.org/10.1016/j.ijhydene.2021.04.050.
[87] Baschuk JJ, Li X. Carbon monoxide poisoning of proton exchange membrane fuel cells. *Int J Energy Res* 2001;25:695–713. https://doi.org/10.1002/er.713.
[88] Gang X, Qingfeng L, Hjuler HA, Bjerrum NJ. Hydrogen oxidation on gas diffusion electrodes for phosphoric acid fuel cells in the presence of carbon monoxide and oxygen. *J Electrochem Soc* 1995;142:2890–3. https://doi.org/10.1149/1.2048660.
[89] Kaserer S, Rakousky C, Melke J, Roth C. Design of a reference electrode for high-temperature PEM fuel cells. *J Appl Electrochem* 2013;43:1069–78. https://doi.org/10.1007/s10800-013-0567-9.

[90] Hinds G, Brightman E. In situ mapping of electrode potential in a PEM fuel cell. *Electrochem Commun* 2012;17:26–9. https://doi.org/10.1016/j.elecom.2012.01.007.
[91] Chung DY, Kim H, Chung Y-H, Lee MJ, Yoo SJ, Bokare AD, et al. Inhibition of CO poisoning on Pt catalyst coupled with the reduction of toxic hexavalent chromium in a dual-functional fuel cell. *Sci Rep* 2014;4:7450. https://doi.org/10.1038/srep07450.
[92] Srinivasan S. *Fuel cells: From fundamentals to applications.* Springer Science & Business Media; 2006.
[93] Staffell I. Zero carbon infinite COP heat from fuel cell CHP. *Appl Energy* 2015;147: 373–85. https://doi.org/10.1016/j.apenergy.2015.02.089.
[94] Staffell I, Scamman D, Velazquez Abad A, Balcombe P, Dodds PE, Ekins P, et al. The role of hydrogen and fuel cells in the global energy system. *Energy Environ Sci* 2019;12:463–91. https://doi.org/10.1039/C8EE01157E.
[95] Stolten D, Samsun RC, Garland N, editors. *Fuel cells: Data, facts, and figures.* John Wiley & Sons; 2016.
[96] www.fuelcelltoday.com accessed on 25 June, 2025.

9 Application and future prospects of fuel cells

9.1 APPLICATION OF PEM

Polymer electrolyte membrane fuel cells have emerged as a potential energy generation devices due to combination of many attractive advantages such as low operation temperature, high current density, light weight, durability, and fast start up capabilities. The cost of the fuel cells can be made affordable with potential development of new materials to be utilized as membranes, catalysts, and other parts of the device. PEMFCs find applications in various sectors, starting from small scale to medium scale to large scale [1, 2]. They have potential application in the field of stationary and portable devices having power requirement of less than 10 Kw [3–5]. They can be utilized in light- to heavy-duty power supply and vehicles as well. PEMFCs have been successfully used for a power range of 10–100 kW in automobile engines and transportation trucks. PEMFCs are slowly entering into the sectors such as heavy-duty transport and airplanes with power requirement of greater than 100 kW [6]. Fuel cell ecosystem is conceptually represented in the Figure 9.1. Fuel cell ecosystem runs in four layers. The first layer involves in the generation of power from different energy resources such as fossil fuels, renewable resources, and biogas.

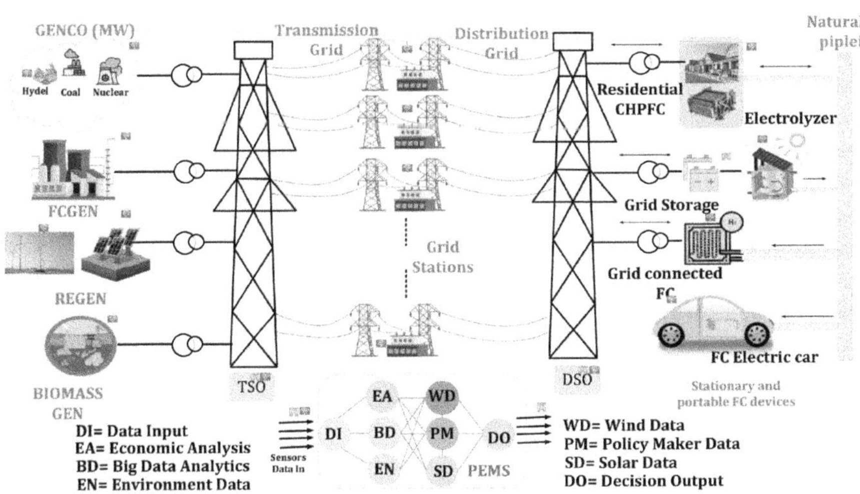

FIGURE 9.1 Fuel cell ecosystem Reprinted with permission from [7]

DOI: 10.1201/9781003733744-9

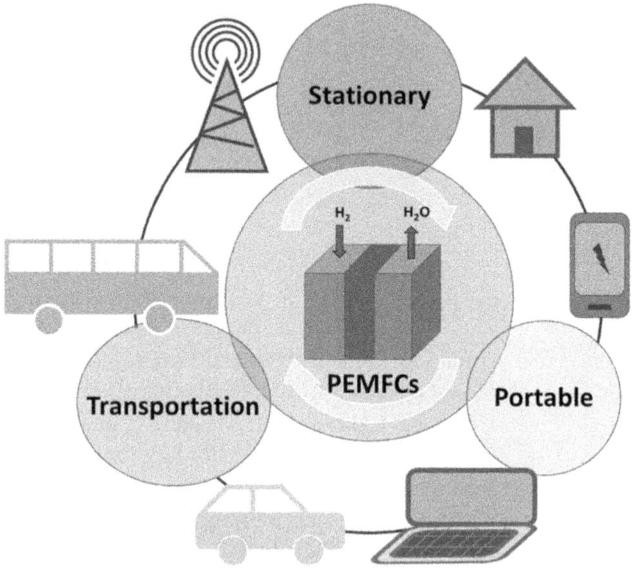

FIGURE 9.2 Application of PEMFCs in different sectors

The power so generated is transferred using grids at grid stations. The second layer is responsible for conversion of voltage and transfer of power from the transmission grid to distribution grid. The third layer is known as consumer layer where various kinds of fuel cells exist.

There are being explored by various companies, e.g., Ballard, Toshiba, GE, Hyundai, Nissan, Renault, Ford, IBM, Samsung, BMW, and many more at the experimental and commercial stages as new technologies to be utilized worldwide. Many of the companies are exploring and continuing their research and expected to grab a large portion of the global market in near future. In addition to stationary, portable, transport, power sources, etc., PEMFCs have also contributed to the production of high valued chemicals. The production of chemicals using the new technology of fuel cells ensures many advantages over the traditional methods. For example, the continuous supply of reactants along the electrodes enhances the catalytic performances. In addition, the isolation of reactant using fuel cell stakes will enhance system safety with ease of separation of products from the reactants in the system. Again, utilization of fuel cells for production of various chemicals along with generation of power is a new approach contributing to the sustainability and energy utilization in an environmentally friendly way. The various application sectors of PEMFC are represented in Figure 9.2.

9.1.1 Transportation using PEMFCs

Among various fields of applications, PEMFCs are most promising and economical for the transportation sector. PEM fuel cells are widely considered as a clean alternative for powering vehicles, particularly in hydrogen-powered cars, buses, trucks, and even trains [8, 9]. They offer fast refueling times and long driving ranges, enabling them ideal for long-distance travel. The utilization of fuel cells for transportation purpose requires a successful combination of fuel cell devices along with an efficient storage technology with heat and water management techniques. Based on this concept the electric vehicles powered by modern fuel cells can be categorized in different types such as all fuel vehicles, hybrid vehicles, passive systems, and active systems. In the case of all power types of vehicles, the power is completely supplied by fuel cell itself. However, such devices lack in power density and power responses that limit their further commercialization. In hybrid types, the major shortcoming of all fuel cell vehicle is resolved and this particular type works in combination with integrated energy storage facilities such as batteries or supercapacitors [10]. For hybrid vehicles, the baseline requirement of power is supplied by fuel cells whereas kickoff power, peak power demand, or any kind of dynamic power demand is supplied by the energy storage devices. The energy storage devices are further charged or discharged depending on the power demand and supply chain. This kind of systems is the most successful one and may operate with highest efficiency in practical scenario.

Passive system is observed in the case of DC bus operated with full cells. In this case, fuel cells are connected directly with a battery and the voltage output is coupled. This system is cost efficient due to omission of DC/DC converter that is generally incorporated after the fuel cells. In contrast, for active system, DC/DC converter is incorporated to gradually add on the fuel cell voltage to the battery system. This provides better control over the voltage output as per power demand. Similar type of systems is also developed with supercapacitors where kick off power requirement is more. Although use of hybrid system makes the operation a little complex and costly however, provides the added advantage of better control and flexibility over the other system with better efficacy.

9.1.2 Stationary power sector

Small scaled power generation occurs by distribution of electrical energy to nearby users. In current era, need of energy generation from renewable sources opens up an attractive path for small-scale stationary power supply. For these applications, PEMFCs are successfully utilized in residential cogeneration systems. PEM fuel cells can be used in both residential and commercial applications for backup power, off-grid power, and as a primary energy source (Figure 9.3). Due to technical challenges associated with production and storage of pure hydrogen, the application of PEMFCs in the stationary power sector needs to be explored further. At present, MCFC and SOFC are considered to be the best for stationary application instead of their low energy efficacy. The main reason behind this is the easy availability of pure fuels for these fuel cells as compared to hydrogen-based fuel cells. Due to these practical concerns, PEMFCs are more explored for applications requiring small power range, generally 1–5 kW and extended to middle or large powered vehicles using hybrid devices. However, high efficiency of PEMFCs is expected once hydrogen production technologies become more efficient. Wang et al. developed components for a stationary power generator [11]. The device comprised of a 5 kW stack made of 56 cells. The MEA is fabricated with a Nafion membrane with platinum as the

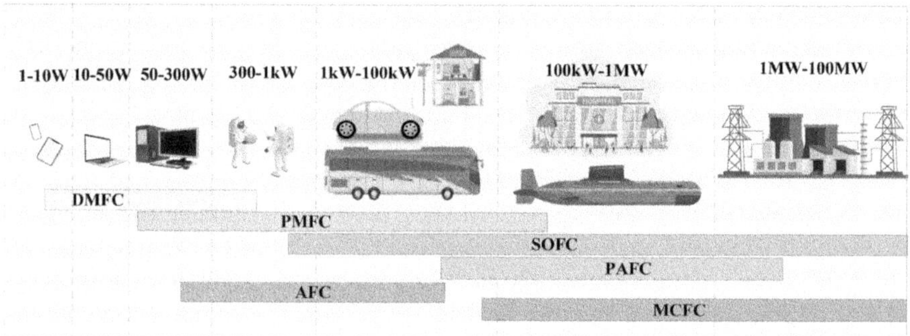

FIGURE 9.3 Portable and stationary application of all fuel cells Reprinted with permission from [7]

catalyst. The 56 individual cells are combined using bipolar plates made of graphite. The device was also supported with external humidification and heat management systems for better performance.

PEMFCs are also utilized effectively in uninterrupted power systems by providing power at small scale to the remote areas when power grid is non-functional. In many remote locations of Asian countries, PEMFC stacks are used for two wheelers [12]. PEMFCs of ≤10 kW is used for making handling trucks such as pallet trucks and forklifts. In Japan about 150 units have been utilized in the airport and by 2030 it is expected to reach 10000 units of frock lift made of fuel cells [13].

9.1.3 Portable Power Generators

Due to their compact size and efficiency, PEM fuel cells are also being explored for use in portable electronics, such as laptops, smartphones, and other battery-powered devices. Portable power generators are mainly used for emergency and personal usages. In Singapore, Horizon Fuel cell Technology made fuel cell stacks to be utilized in medical devices [14]. Other companies worldwide, like Intelligent Energy and Power UP Energy Technologies, producing efficient power using PEMFCs for small-scale power generators [15, 16]. In militaries, PEMFCs have potential to replace conventional heavy power system that is impractical to be carried by the soldiers to fulfill the requirement of power. Small-scale, portable generators can be of effective choice in these cases for weight efficient power supply systems. In India, Naval Research Center focuses on the development of 100 W power for 10 consecutive hours for the soldiers using fuel cell technologies [17]. PEM fuel cells are utilized for backup power in energy uses such as in hospitals, data centers, and telecommunication networks, providing clean, reliable energy when the grid goes down [18, 19].

9.1.4 Production of Chemicals

PEMFCs in general are utilized for the production of noise free electricity with high power densities and can also be used to provide easy reaction condition for production of chemicals with good selectivity [20]. Study suggests that the PEMFCs can be

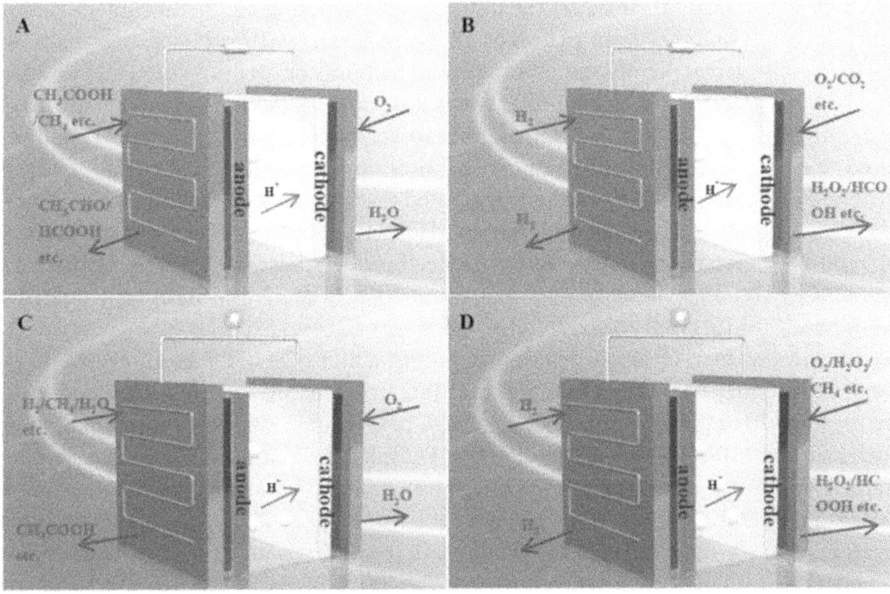

FIGURE 9.4 Representation of simultaneous generation of electricity and chemicals using PEM fuel cells Reprinted with permission from [23]

utilized as reactors for synthesis of various chemicals where hydrogenation proceeds at the cathode and materials with oxidation ability, e.g., oxygen, CO_2, some ketones can be used as cathode [21]. When overall enthalpy of the device becomes positive, a partial voltage is applied for the reaction to proceed. Again, if the free enthalpy change is negative then chemical and electricity both can be produced simultaneously [22, 23]. The study reveals that various reactions taken place at the electrodes during the process. At anode, alcoholic oxidation, methane oxidation, and propane epoxidation are observed [22]. At cathode side, main reactions that are evidenced include hydrogenation of alkynes, peroxide production, and reduction of CO_2. However, as per literature, the value added chemicals cannot be produced simultaneously both at the cathode and anode side. In each case, platinum or palladium over carbon support is utilized as catalyst for the reaction to proceed. The chemicals so generated are of good selectivity and quality; however, the yield is quite low that demands further process optimization before commercialization of this technology [24–26]. Figure 9.4 represents simultaneous generation of electricity with chemicals using PEM fuel cells

9.2 INDIAN AND GLOBAL SCENARIO OF PEMFC AND ASSOCIATED CHALLENGES

Currently in India and around the globe, PEMFCs are grabbing the above-mentioned market gracefully where they are used mainly for the generation of electricity and production of valuable chemicals, e.g., ethanol, propane, and formic acid. The research is primarily focused on the development of non-noble metal catalysts, specifically catalysts other than platinum for cost efficiency. Efforts are also made to develop commercial membranes alternative to Nafion® or perfluroniated membranes that can serve similar to Nafion® at much low cost. Although chemicals are produced by different industries using PEM however, some challenges are still encountered, e.g., limited option of catalytic reaction modification, restriction in catalyst type and cost, lower rate of conversion, limited research data on reaction kinetics and chemical pathways. Hyundai, Toyota, and Honda are the key players in the market of light-duty electric vehicles run by fuel cells. The statistical analysis of the sold vehicles of Hyundai Nexo reveals that about 63% of their sold light duty vehicles were based on fuel cell technology in the year of 2019 [27]. The other companies leading in the market are Toyota contributing about 32%, Honda about 5% [27]. Further Hyundai has a target of 500,000 fuel cell electric vehicles manufacturing by the year 2030 [28]. In Europe, Peugeot, Opel, etc., are working on production of fuel cell van [1]. However, company like Daimler discontinued their GLC hybrid SUV from 2018 and close project on hydrogen car manufacturing [29]. Zero emission heavy-duty vehicles are also in recent trends. In the year 2023–2024, about twelve models of heavy-duty vehicles are available in the global market and eight additional is underway [30]. Research for commercialization is focused on the manufacturing of heavy-duty vehicles for public transport. About 12 European bus original equipment manufacturer and available for fuel cells driven buses and it is expected to be gaining hugely to about 1300 in upcoming years [31]. Fuel cells markets are also growing in material handling and machines [32]. Light-weight vehicles and electrically powered bicycle market have also shown potential in recent past [33]. In Taiwan and China, it is utilized as a source of daily communication purpose. In 2005, leisure yachts are developed of 300 W PEMFC system. These are focused for application in sailing yachts since the ingestion of power is limited during long cruises due to small battery backup [34]. Under these circumstances, a noiseless environmentally friendly fuel cell technology ensures to meet demand of energy at the time of non-availability of conventional resources.

9.3 OUTLOOK AND FUTURE PROSPECTIVE OF PEMFC

In general, the performance evaluation of a fuel cell stack depends on four main factors: performance, cost of operation, durability, and physical factors [35]. The performance factor of a fuel cell may include efficiency, polarization, and power density. These factors can be regulated by optimization of stack design to achieve maximum possible output. Cost of operation can be determined by factors, e.g., system management, thermal and water, fuel consumption, parasitic load on the system, and efficiency of power conditioning instruments. Durability as understood is a general factor that demands prolonged operation with minimal or no maintenance specifically those intended to be used for stationary applications. Finally, the evaluation factor of a fuel cell stack is dependent on various physical parameters, such as stack-specific parameters, size, weight, numbers; the number of cells in operation; cell active area; and total system weight. For the fuel cell intended to be operated for transport applications, the weight and size of a fuel cell become a critical factor under

TABLE 9.1
DOE target for membrane to be used for transport applications [31, 36]

Parameters	Units	Year 2015	Year 2025
Maximum H_2 crossover	mA cm^{-2}	2.4	2
Maximum O_2 crossover	mA cm^{-2}	1,1	2
Area specific proton resistance	Ohm cm^{-2}	0.027 (25 kPa)	0.02
Maximum operating temperature	°C	120	120
Minimum electrical resistance	Ohm cm^{-2}	>5600	1000
Durability (mechanical and chemical)	h	>5000	>8000
Cost	$/kW	53	30

consideration. Table 9.1 summarizes the DOE target for the membrane material to be used in transport applications.

Fuel cells are the environmentally friendly energy conversion devices of the modern age with several features, such as low harmful emission, highly efficient, dynamic load following, modular, and static silent devices, which open up their suitability for numerous applications, ranging from micro-fuel cells to multi-Mega Watt giant power plants. In the recent decade, with the increasing environmental concern of the world leaders, research on the suitable fuel cell materials grabs the attention of the researchers. Fuel cells justified their dignity and superiority as compared to other energy conversion devices. However, major challenges which confine their wide acceptance are high cost, low durability, problem in heat and thermal management, need of an expensive platinum catalyst, sensitiveness toward contamination as well as use of hydrogen as a source of fuel that needs special, dedicated infrastructure for production and storage as well as simultaneous emergence of the similar energy devices. These remain the main impediments toward the successful development and commercialization of the fuel cell devices.

Cost and durability are the two major impediments toward the world-wide commercialization of PEMFCs. Development of the low cost and self-healing PEM, capable of recovering the loss of conductivity during long operation, would be an interesting attempt in this direction. The total cost of the device could be reduced by judicial selection and the innovative design of the alternative membranes and using efficient catalysts. The alternative membranes may be of close in conductivity as that of Nafion®, however, they must be of high durability, dimensionally stable, and should possess low fuel cross over and low cost of production as well as operation.

From the property comparison of various polymeric membrane materials as shown in Figure 9.5, it is concluded that no single PEM material would be able to work successfully in diverse application areas such as automotive and portable devices. However, what needed is that the understanding of the actual requirement of the system to select an appropriate polymer electrolyte membrane for the intended

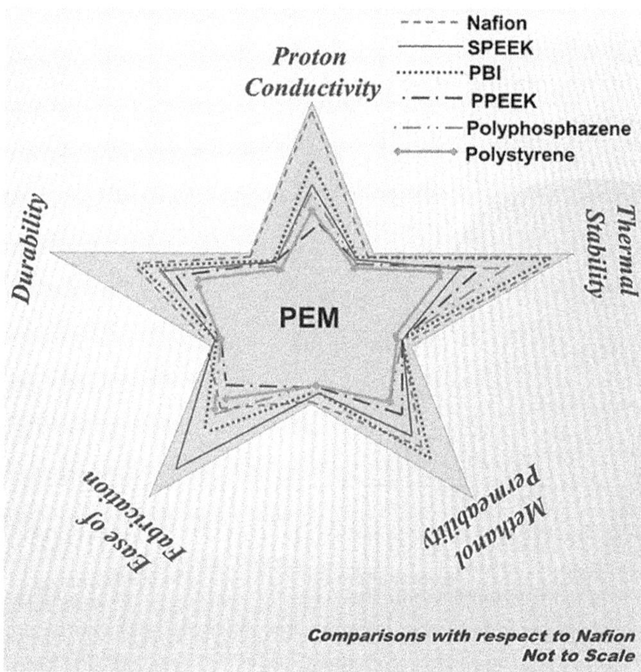

FIGURE 9.5 Property comparison of some important class of alternative PEM to the commercial Nafion® by considering the progress of alternative membranes from the literature

application. The current research trends on the innovative membrane materials and associated challenges that would be a guideline for the researchers working in this field.

9.4 CHAPTER SUMMARY

The success of fuel cells in various application sectors no doubt depends on the supply of highly pure hydrogen at affordable cost and required safety protocol. At current scenario, the scope of PEMFC/Battery/Supercapacitor devices in integration may be capable to compete with the traditional power sources due to multiple advantages. PEMFCs have also potential to be utilized as effective reactor for the production of chemicals and electricity both. The slow response of PEMFCs as per power requirement needs the development of light-weight hybrid power devices capable to generate large power promptly. A proper management of energy storage devices in combination of fuel cells would provide efficiency and dynamic power demand. The success of the fuel cells is again dependent on the efficient hydrogen storage system capable to provide highly pure hydrogen. The hydrogen generation and storage for small scale and large scale commercial utilization of fuel cell devices would need a compressed gas storage facility and or metal storage system in high volumetric capacities with robust safety precaution strategies. This would have been beneficial for the development of a complete PEMFC based power generation system in compact and space efficient manner. We hope, with the present passion of research on PEMs, a bright and successful near future is awaiting with huge commercialization of hydrogen-powered devices.

REFERENCES

[1] Selmi T, Khadhraoui A, Cherif A. Fuel cell–based electric vehicles technologies and challenges. *Environ Sci Pollut Res* 2022;29:78121–31. https://doi.org/10.1007/s11356-022-23171-w.
[2] Wang Y, Ruiz Diaz DF, Chen KS, Wang Z, Adroher XC. Materials, technological status, and fundamentals of PEM fuel cells – A review. *Mater Today* 2020;32:178–203. https://doi.org/10.1016/j.mattod.2019.06.005.
[3] Wang Z, Liu Z, Fan L, Du Q, Jiao K. Application progress of small-scale proton exchange membrane fuel cell. *Energy Rev* 2023;2:100017. https://doi.org/10.1016/j.enrev.2023.100017.
[4] Mancino A, Menale C, Vellucci F, Pasquali M, Bubbico R. PEM fuel cell applications in road transport. *Energies (Basel)* 2023;16:6129. https://doi.org/10.3390/en16176129.
[5] Wee J-H. Applications of proton exchange membrane fuel cell systems. *Renewable Sustainable Energy Rev* 2007;11:1720–38. https://doi.org/10.1016/j.rser.2006.01.005.
[6] Cullen DA, Neyerlin KC, Ahluwalia RK, Mukundan R, More KL, Borup RL, et al. New roads and challenges for fuel cells in heavy-duty transportation. *Nat Energy* 2021;6:462–74. https://doi.org/10.1038/s41560-021-00775-z.
[7] Sebbani I, Ettouhami MK, Boulakhbar M. Fuel cells: A technical, environmental, and economic outlook. *Cleaner Energy Syst* 2025;10:100168. https://doi.org/10.1016/j.cles.2024.100168.
[8] Hwang JJ, Wang DY, Shih NC, Lai DY, Chen CK. Development of fuel-cell-powered electric bicycle. *J Power Sources* 2004;133:223–8. https://doi.org/10.1016/j.jpowsour.2004.02.004.
[9] Hwang J, Wang D, Shih N. Development of a lightweight fuel cell vehicle. *J Power Sources* 2005;141:108–15. https://doi.org/10.1016/j.jpowsour.2004.08.056.
[10] Verma S, Mishra S, Gaur A, Chowdhury S, Mohapatra S, Dwivedi G, et al. A comprehensive review on energy storage in hybrid electric vehicle. *J Traffic Transp Eng (English Edition)* 2021;8:621–37. https://doi.org/10.1016/j.jtte.2021.09.001.
[11] Wang C, Mao Z, Bao F, Li X, Xie X. Development and performance of 5kw proton exchange membrane fuel cell stationary power system. *Int J Hydrogen Energy* 2005;30:1031–4. https://doi.org/10.1016/j.ijhydene.2004.11.010.
[12] Assaf J, Shabani B. Economic analysis and assessment of a standalone solar-hydrogen combined heat and power system integrated with solar-thermal collectors. *Int J Hydrogen Energy* 2016;41:18389–404. https://doi.org/10.1016/j.ijhydene.2016.08.117.
[13] Hydrogen and Fuel Cell Strategy Council, *The strategic road map for hydrogen and fuel cells*. IGEM: Institution of Gas Engineers and Managers (IGEM); 2019.
[14] Horizon Fuel Cell Technologies Inc., Fuel cell systems, 2022. Https://WwwHorizonfuelcellCom/Fuelcellsystems.
[15] Intelligent energy's 2.4kW fuel cell power module integrated into latest UAV product from ISS aerospace. Intelligent Energy Inc, Lightweight and Compact Fuel Cell Power; 2020.
[16] *The world's first portable hydrogen fuel cell based smart generator.* Powerup Energy Technologies; 2022.
[17] Narayana Das J. *Fuel cell technologies for defence applications.* Energy Engineering, Singapore: Springer Singapore; 2017, p. 9–18. https://doi.org/10.1007/978-981-10-3102-1_2.
[18] Staffell I, Scamman D, Velazquez Abad A, Balcombe P, Dodds PE, Ekins P, et al. The role of hydrogen and fuel cells in the global energy system. *Energy Environ Sci* 2019;12:463–91. https://doi.org/10.1039/C8EE01157E.

[19] Sorlei I-S, Bizon N, Thounthong P, Varlam M, Carcadea E, Culcer M, et al. Fuel cell electric vehicles—a brief review of current topologies and energy management strategies. *Energies (Basel)* 2021;14:252. https://doi.org/10.3390/en14010252.

[20] Rodríguez-Gómez A, Dorado F, de Lucas-Consuegra A, de la Osa AR. Influence of Pt/Ru anodic ratio on the valorization of ethanol by PEM electrocatalytic reforming towards value-added products. *J Energy Chem* 2021;56:264–75. https://doi.org/10.1016/j.jechem.2020.07.061.

[21] Iguchi S, Kataoka M, Hoshino R, Yamanaka I. Direct epoxidation of propylene with water at a PtO_x anode using a solid-polymer-electrolyte electrolysis cell. *Catal Sci Technol* 2022;12:469–73. https://doi.org/10.1039/D1CY01888D.

[22] Lee B, Hibino T. Efficient and selective formation of methanol from methane in a fuel cell-type reactor. *J Catal* 2011;279:233–40. https://doi.org/10.1016/j.jcat.2010.12.020.

[23] Jiang X, Chen R, Chen Y-X, Lu C-Z. Proton exchange membrane fuel cells: Application for value-added chemical productions. *Chem Synth* 2024;4. https://doi.org/10.20517/cs.2023.44.

[24] Rodríguez-Gómez A, Dorado F, Sánchez P, de la Osa AR. Boosting hydrogen and chemicals production through ethanol electro-reforming on Pt-transition metal anodes. *J Energy Chem* 2022;70:394–406. https://doi.org/10.1016/j.jechem.2022.02.028.

[25] Serrano-Jiménez J, de la Osa AR, Rodríguez-Gómez A, Sánchez P, Romero A, de Lucas-Consuegra A. Graphene-like materials as an alternative to carbon Vulcan support for the electrochemical reforming of ethanol: Towards a complete optimization of the anodic catalyst. *J Electroanal Chem* 2022;921:116680. https://doi.org/10.1016/j.jelechem.2022.116680.

[26] Abdelnasser S, Hakamata T, Ogihara H, Kurokawa H. Electrochemical oxidation of 1-propanol through proton exchange membrane electrolysis. *J Electroanal Chem* 2023;928:117009. https://doi.org/10.1016/j.jelechem.2022.117009.

[27] Kane M. Hydrogen fuel cell car sales in 2019 improved to 7,500 globally. 2020.

[28] Hyundai News. Hyundai motor group reveals 'FCEV Vision 2030'. 2018.

[29] Automotive News Europe. Daimler Will end development of fuel cell cars, 2020. Https://EuropeAutonewsCom/Automakers/Daimler-Will-End-Development-Fuel-Cell-Cars.

[30] IEA. Trends and developments in EV markets. In *Global EV Outlook 2023. Catching up with Climate Ambitions*; IEA; 2023, France.

[31] Wang Y, Seo B, Wang B, Zamel N, Jiao K, Adroher XC. Fundamentals, materials, and machine learning of polymer electrolyte membrane fuel cell technology. *Energy and AI* 2020;1:100014. https://doi.org/10.1016/j.egyai.2020.100014.

[32] Yartys VA, Lototskyy M V., Linkov V, Pasupathi S, Davids MW, Tolj I, et al. HYDRIDE4MOBILITY: An EU HORIZON 2020 project on hydrogen powered fuel cell utility vehicles using metal hydrides in hydrogen storage and refuelling systems. *Int J Hydrogen Energy* 2021;46:35896–909. https://doi.org/10.1016/j.ijhydene.2021.01.190.

[33] Li M, Bai Y, Zhang C, Song Y, Jiang S, Grouset D, et al. Review on the research of hydrogen storage system fast refueling in fuel cell vehicle. *Int J Hydrogen Energy* 2019;44:10677–93. https://doi.org/10.1016/j.ijhydene.2019.02.208.

[34] Beckhaus P, Dokupil M, Heinzel A, Souzani S, Spitta C. On-board fuel cell power supply for sailing yachts. *J Power Sources* 2005;145:639–43. https://doi.org/10.1016/j.jpowsour.2004.12.078.

[35] Folkesson A, Andersson C, Alvfors P, Alaküla M, Overgaard L. Real life testing of a Hybrid PEM Fuel Cell Bus. *J Power Sources* 2003;118:349–57. https://doi.org/10.1016/S0378-7753(03)00086-7.

[36] Satyapal S. U.S. department of energy hydrogen and fuel cell technology perspectives. 2019. https://doi.org/10.2172/1511435.

Index

Note: **Bold** page numbers refer to tables and *italic* page numbers refer to figures.

R

S

T

For Product Safety Concerns and Information please contact our EU
representative GPSR@taylorandfrancis.com
Taylor & Francis Verlag GmbH, Kaufingerstraße 24, 80331 München, Germany

www.ingramcontent.com/pod-product-compliance
Ingram Content Group UK Ltd.
Pitfield, Milton Keynes, MK11 3LW, UK
UKHW021507240926
475702UK00035B/619
* 9 7 8 1 0 3 2 8 5 4 8 3 0 *